Laboratory Manual

for

Introductory Chemistry
Third Edition

by

Steve Russo
Cornell University

Mike Silver
Hope College

Doris Kimbrough
University of Colorado at Denver

Wendy Gloffke
Cedar Crest College

PEARSON
Benjamin Cummings

San Francisco • Boston • New York
Cape Town • Hong Kong • London • Madrid • Mexico City
Montreal • Munich • Paris • Singapore • Sydney • Tokyo • Toronto

Editor-in-Chief	*Adam Black*
Publisher	*Jim Smith*
Project Editor	*Katherine Brayton*
Editorial Assistants	*Kristin Rose and Grace Joo*
Senior Marketing Manager	*Scott Dustan*
Managing Editor	*Corinne Benson*
Production Supervisor	*Shannon Tozier*
Production Service and Compositor	*Progressive Publishing Alternatives*
Illustrations	*B&B Illustrations*
Manufacturing Buyer	*Pam Augspurger*
Cover Image	*Emi Koike and Quade Paul*
Cover Design	*Progressive Publishing Alternatives*
Printer	*Technical Communications Services*

ISBN 0-8053-0478-9

www.aw-bc.com

1 2 3 4 5 6 7 8 9 10—TCS—09 08 07 06

Additional resources available for students:

Problem Solving Guide and Workbook (0-8053-0548-3)
By Saundra Yancy McGuire of Louisiana State University
Provides over 200 worked examples and more than 550 practice problems and quiz questions to help students develop and practice their problem-solving skills.

Study Guide and Selected Solutions (0-8053-0552-1)
By Saundra Yancy McGuire of Louisiana State University
Features examples from each chapter, learning objectives, review of key concepts from the text, and additional problems for student practice. Also provides comprehensive answers and explanations to selected end-of-chapter problems from the text.

Chemistry Place™ for *Introductory Chemistry, 3rd edition*
www.aw-bc.com/chemplace
The Chemistry Place engages students in interactive exploration of chemistry concepts and provides a wealth of tutorial support. Tailored to Russo and Silver's Third Edition, the site includes detailed objectives for each chapter of the text, interactive tutorials featuring simulations, animations and 3-D visualization tools, multiple-choice and short-answer quizzes, InterAct Math with content correlated to the text for math concepts practice, an extensive set of Web links, and audio preludes for each chapter.

Additional resources available for instructors:

Instructor Teaching Guide and Complete Solutions (0-8053-0409-6)
By Saundra Yancy McGuire of Louisiana State University
Includes chapter summaries, complete descriptions of appropriate chemical demonstrations for lecture, suggestions for addressing common student misconceptions, and examples of everyday applications of selected topics for lecture use, as well as the solutions for all the problems in the text.

Printed Test Bank (0-8053-0448-7)
By Paris Svoronos and Soraya Svoronos of Queensborough Community College of the City University of New York, with contributions by Christopher Truitt of Texas Tech University
This printed test bank includes over 1700 questions that correspond to the major topics in the text.

Computerized Test Bank (0-8053-0549-1)
By Paris Svoronos and Soraya Svoronos of Queensborough Community College of the City University of New York, with contributions by Christopher Truitt of Texas Tech University
This dual-platform CD-ROM includes over 1700 questions that correspond to the major topics in the text.

Instructor Manual for the Laboratory Manual (0-8053-0408-8)
By Wendy Gloffke of Cedar Crest College and Doris Kimbrough of the University of Colorado at Denver

Instructor Resource CD (0-8053-0516-5)
This CD-ROM includes all the art and tables from the book in the best resolution for use in classroom projection or creating study materials and tests. In addition, the Instructor can access the PowerPoint® lecture outlines to highlight key points in his or her lecture. Also available on the IRCD are downloadable files of the Instructor Solution Manual, the Test Bank with more than 1000 questions, and a set of "clicker questions," suitable for use with classroom-response systems.

Transparency Acetates (0-8053-0488-6)
Includes 125 full-color acetate transparencies.

CourseCompass™
CourseCompass™ combines the strength of Benjamin Cummings content with state-of-the-art eLearning tools! Course-Compass™ is a nationally hosted, dynamic, interactive online course management system powered by BlackBoard, leaders in the development of Internet-based learning tools. This easy-to-use and customizable program enables professors to tailor content and functionality to meet individual course needs! Every CourseCompass™ course includes a range of preloaded content such as testing and assessment question pools, chapter-level objectives, chapter summaries, photos, illustrations, videos, animations, and Web activities—all designed to help a student master core course objectives. Visit www.coursecompass.com for more information.

Contents

Preface

This laboratory manual is written to accompany *Introductory Chemistry* by Steve Russo and Mike Silver; however, it can be used with any introductory chemistry text or as a stand-alone resource for the course. As a beginning chemistry student, you are about to embark upon a captivating journey through an invisible world of atoms and molecules. As you get involved in working problems assigned to you from the lecture portion of the class, it is often easy to forget that chemistry is first and foremost a laboratory science. All of the theories, principles, and scientific laws discussed in your textbook were first observed, advanced, and elucidated in the chemistry laboratory. The objective of this laboratory curriculum is to promote an understanding and appreciation for the laboratory component of chemistry, including:

- Developing basic lab skills and familiarity with common laboratory equipment.
- Building an awareness of the safety considerations in a laboratory setting.
- Learning to collect, organize, analyze, and present data effectively.
- Exercising critical thinking.
- Understanding the relationship between macroscopic observations and atomic and molecular properties.

In keeping with the molecular emphasis of the Russo and Silver text, the approach used in this manual will facilitate your understanding of the molecular level activity that produces the macroscopic evidence of chemical or physical phenomena. In addition there is a strong emphasis on the use of common consumer items in the experiments (such as foods, over-the-counter medicines, and household products) so that you, the beginning chemistry student, can develop an appreciation of how chemistry fits into your universe.

The data collection and reporting follows a progression. At the beginning, we have provided ready-made data tables. Pay attention to how they are organized and how they are used to collect, organize, and report the information that you gain from the experiments. As the course progresses and you become a more sophisticated chemist, you will be expected to create your own data tables and graphs.

Each laboratory unit has an introductory section that provides the rationale for the experiment followed by a pre-lab exercise. These pre-lab exercises are designed to promote understanding of the procedure, and provide an opportunity to practice any calculations or other determinations that you will need to understand and report your data. Resist the temptation to show up at lab unprepared. Reading each laboratory unit and working through the pre-lab assignment before coming to the the lab, will save you much time and aggravation while doing the experiment. As you read each unit, visualize what you will do, try to anticipate what data are to be collected, then consider the best method for organizing those data. Keep in mind the rationale behind each step in the experiment and what your data mean within the context of the principles you learn in lecture.

Specific Laboratory Skills

The laboratory units in the manual will give you opportunities to acquire basic laboratory skills, including:

- Measuring volumes and masses
- Using common glassware
- Preparing solutions
- Titration
- Filtration
- Use of precipitation reactions to identify unknown salt solutions
- Chromatography
- Use of the Spectronic-20
- Use of physical and chemical properties to identify unknowns
- Experimental design

Acknowledgments

I would like to first and foremost acknowledge the help and support of my family for support, comfort, and enthusiasm for all my projects. I would also like to thank Jimmy Reeves at the University of North Carolina at Wilmington and my colleagues at Colorado University at Denver (and at Purdue) for professional support and guidance. I must also mention the contributions of Lydia Walsh, Craig Swank, Kim Henry, and Daniel Haun in laboratory development and procedure editing, as well as all the CU-Denver students who were willing and enthusiastic about trying new experiments. Also thanks to the folks at Benjamin/ Cummings for hanging in there with me through all the upheavals!

Doris Kimbrough
University of Colorado at Denver

I must acknowledge the generosity of Steve Russo from Cornell and Gene Reck from Wayne State University for sharing labs that they developed for their classes. I would also like to thank the following reviewers for their valuable comments and contributions:

Walter Dean, *Lawrence Technological University*

Donna Friedman, *St. Louis Community College*

John Goodwin, *Coastal Carolina University*

The team at Benjamin/Cummings is wonderful and I am grateful for their encouragement. Robin J. Heyden, Ben Roberts, Claudia Herman, Emi Koike, and Margot Otway each provided guidance in the design and execution of this project.

Finally, one needs sustenance, both physical and emotional, when completing a project of this nature. My husband, Thom Kotch, cheerfully provided gourmet meals and a supportive environment so that I could work without distraction. I truly couldn't have done it without him.

Wendy Gloffke
Cedar Crest College

Lab Safety: In Case of Emergency

OBJECTIVES

- Learn the nature and location of safety features in your lab.
- Learn how to use the safety equipment in the lab.
- Understand proper lab protocol to ensure safe lab experiences.

EQUIPMENT/MATERIALS

Safety goggles, list of safety features specific to your laboratory.

INTRODUCTION

What would you do if a fire broke out in the lab? What if your lab partner spilled a large quantity of dilute acid on herself—how should you respond? Situations such as these rarely occur in a chemistry lab, but it's important to know what to do should the need arise. The greatest contribution you can make to lab safety is being aware of any hazards associated with what you are doing before you start working and during your experiment. This means *reading the lab before you start* and paying attention to what you are doing at all times. Any safety precautions are noted in the procedure part of each lab, and an initial run-through of the experiment in your head will alert you to procedures, reagents, equipment, and so on, that may require special attention.

Emergency Phone and Phone Numbers

Your instructor will indicate the location of emergency phones and contact numbers. You should also note the locations of other phones near the laboratory.

Emergency phone number: _____

Location of emergency phone: _____

Locations of other phones in the area of the laboratory: _____

_____

Exits

An important aspect of lab safety is knowing how to leave the area quickly and safely. There may be doorways in your lab that lead to closed rooms and should therefore not be used as exits. Some exits may lead directly to an outside door, while others lead to different areas of the building. As you survey the lab and surrounding rooms and corridors, answer the following questions.

Questions

1. Where are the exits in your lab?
2. Where does each exit lead?
3. Which exits are closest to your area?

Fires and Fire Safety Equipment

There are several sources of fire hazards in a chemistry lab. The most common is inattention on the part of a student. Bunsen burners, hot plates, and heating mantles can all become very hot when in use and therefore need to be monitored closely. The open flame of a Bunsen burner needs to be kept away from all flammable material, including papers, chemicals, hair, skin, and clothing. The bench area in which you are working should contain only your experimental equipment, notebook, and text. Keep the working area clear of clutter, especially crumpled papers and paper towels.

Chemicals sometimes ignite in the course of an experiment. You can generally smother small fires by covering them with a watch glass, crucible lid, or some other device. Alert your instructor immediately if you notice smoke, a burning odor, or fire.

Electrical wires and equipment should be kept away from water and flammable materials. Sometimes electrical outlets are positioned in such a way that it is difficult to keep a lot of space between wet areas and wires. Under these circumstances, you must watch your equipment carefully and keep electrical devices turned off and unplugged when not in use. Roll or wrap wires so that they are not sprawled all over the bench area.

Fire extinguishers are marked as to their contents and the types of fires that they will extinguish. Water extinguishers may deliver a mist or a directed spray. Foam extinguishers smother a fire by cutting off the supply of oxygen. Dry chemical and carbon dioxide extinguishers are used to fight specific types of fires. You should acquaint yourself with each type available in your lab in case you need to supply the information to someone. If you are not trained to use an extinguisher, you should not try to put out a fire with one. Large fires require immediate evacuation; leave the fire fighting to professionals.

Questions

1. Is there a fire alarm in the lab or nearby?
2. Where are the fire extinguishers located in the lab?
3. Are any fire extinguishers located outside the laboratory but close by?
4. What types of fire extinguishers are available?
5. Under what circumstances should each be used?

Fire Blankets

Fire blankets provide a surface that is wrapped around a person whose clothing has caught fire. The blanket cuts off oxygen to the fire and smothers it.

Questions

1. Where are fire blankets located?
2. Draw a diagram to show the proper use of a fire blanket.
3. Under what circumstances should a fire blanket be used?

Safety Shower

Safety showers provide copious amounts of water to help rid a large surface area of chemicals. For example, if a large part of your body has been drenched with acid (accidentally, of course), the safety shower is the place for you. Ideally, the person who is put under a safety shower should also be stripped of clothing (no modesty under these circumstances!).

Questions

1. Where are the safety showers located?
2. How are the safety showers used?

Eyewash

Eyewashes are specifically designed to deliver a stream of water directly into the eyes when the head is held in the proper position. There's a catch to using the eyewash, though—you usually need help. If chemicals get into your eyes, grab the nearest warm body and have him run with you to the eyewash. Similarly, if your lab partner or neighbor shrieks and grabs his eyes, don't ask questions: grab him and haul him over to the eyewash. One of you will have to hold your eyelids open for several minutes while the water washes out any irritants.

Questions

1. Where are the eyewashes located?
2. How are the eyewashes used?

Fume Hoods

Fume hoods keep volatile chemicals from building up in the air around you. Whenever a reagent bottle is placed in the hood, it is safe to assume that it should remain there and that all deliveries from that bottle should take place in the hood. Most hoods have water, gas, fan, and light connections. The fan should always be on when there are chemicals in the fume hood.

Questions

1. Where are the fume hoods located?
2. What is the lab policy concerning fume hoods being turned off?
3. How is equipment organized in the hood?
4. Is there a hood devoted to waste disposal?
5. Are there water and gas hookups in the hood?
6. How do you test to see if the hood is working?

Goggles

You should never need to use the eyewash because you should *always* have your safety goggles on in the lab. Safety goggles protect your eyes from accidental chemical splashes, from glass fragments that fly when glassware is dropped and shatters, from misdirected water, and from a whole host of other fairly common mishaps. Your instructor will direct you to sources for lab goggles. Most college bookstores carry them, and they can also be purchased at hardware stores. Make sure you know what type of goggle is acceptable for use in your lab before you purchase a pair.

It is common practice to wear goggles over eyeglasses. Safety policies concerning contact lenses vary. Some laboratory institutions prohibit the wearing of contact lenses under the theory that they can act as traps for chemicals and hold them against the eye, causing damage. However, some studies have indicated that contact lenses actually can *protect* the eye in the event of a splash or spill. If you wear contacts, ask your instructor what your institution's policy is. Goggles should be worn regardless of whether you are wearing contacts.

Questions

1. Under what circumstances should goggles be worn?
2. What purpose do goggles serve?

Chemicals in the Lab

Any specific safety issues are addressed in each lab. You will be alerted when chemicals are corrosive, cause discoloration, are flammable, or present health hazards. Your instructor may have additional safety concerns or instructions.

Acids and bases are common classes of chemicals that are used in every chemistry lab. When mixed together, acids and bases neutralize each other. The acids and bases with which you will be working are usually dilute. This means that there are small numbers of acidic or basic molecules in a large amount of water. Dilute acids and bases can cause minor skin irritation, eye damage, or material corrosion. If any acid is spilled, a neutralizing base such as sodium bicarbonate (baking soda) should be mixed with the acid. If a base is spilled, a neutralizing acid such as acetic acid (vinegar) should be mixed with the base.

When you need to add water to an acidic solution, *the acid should be added to the water*, not the other way around. Adding water to acid may cause spattering to occur. The reaction is also very exothermic, which causes the container to get warm. Remember the following when adding acid and water: "Do as you oughta: add acid to watta!"

Disposal

At some time you may need to dispose of broken glass, broken thermometers, or chemical waste. Broken glass that isn't chemically contaminated is usually put into a designated container. It's important that you do not put broken glassware in with regular trash because of the hazard it can pose to janitorial staff when they collect it. When a mercury-containing thermometer breaks, it must be collected and deposited into a specified waste container. Most labs have mercury collection kits available. Do not put broken mercury thermometers in the glass disposal unit.

Chemicals such as solvents, organic waste, heavy metals, and so on, must be disposed of as prescribed by federal safety guidelines. Your instructor will outline the appropriate disposal guidelines when the need arises.

Questions

1. Where is the broken-glass disposal container located? Does it look different from other disposal containers?
2. Is there a broom, dustpan, and brush near the glass disposal container? If not, where can you find these items?
3. What is the procedure for disposal of a broken mercury thermometer?
4. Is there a designated area for chemical waste disposal?

Personal Items in the Lab

Your instructor will direct you in how to store personal items (purses, backpacks, jackets, etc.) while working in the laboratory. They should not be left on the bench or scattered around the floor where they present hazards for spills, fires, or tripping your fellow students. In addition, you should never have food or drink in the chemistry laboratory.

Safety in the Lab: Test Your Understanding

Your instructor may assign groups to discuss the figures on the following three pages and the example stations set up in your laboratory. He or she may also require you to address these situations by writing a few brief sentences.

Consider each of the figures in light of your understanding of the chemistry laboratory and the activities that go on there. Do any of the figures show unsafe lab practices? Explain. How would you respond to each situation?

Examine each of the example stations as directed by your instructor. Discuss each setup with your group, and determine where the dangers and mistakes are.

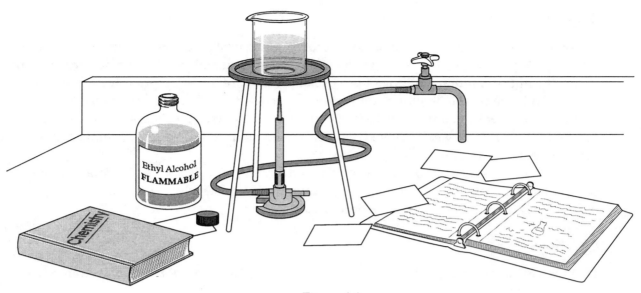

Figure 1.1

Figure 1.2

Figure 1.3

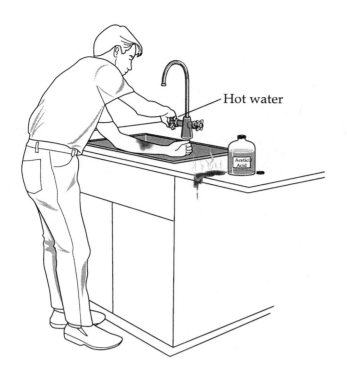

Figure 1.4

Figure 1.5

Figure 1.6

Figure 1.7

Figure 1.8

2

Identifying a Mystery Substance: An Examination of Physical and Chemical Properties

OBJECTIVES

- Identify the physical and chemical properties of substances.
- Learn to record observations in a data table.
- Identify an unknown substance by its properties.

EQUIPMENT/MATERIALS

Sodium chloride, sodium bicarbonate, calcium carbonate, citric acid, distilled water, vinegar, ammonia, red cabbage extract, mineral oil, well plates or test tubes, hand lenses, spatulas, stirring rods or toothpicks, pipets or eyedroppers, unknown solid.

INTRODUCTION

Humans instinctively organize their world in order to interact with it. We group animals and plants into categories according to their appearance, behavior, and other characteristics. When we describe architecture or artistry, we do so by comparing one work to another. We prefer one genre of music or books to another. Scientists organize the subjects of their study—often imposing categories beyond what the area of interest can legitimately sustain.

This organization process is useful because we depend on similar things behaving similarly. You would certainly be surprised if the orange you were eating suddenly had the taste and consistency of a hard-boiled egg, rather than behaving like every other orange you had ever eaten! Furthermore, upon encountering a new fruit that is similar to an orange—say a grapefruit—you might expect it to look, taste, and otherwise behave similarly to an orange, based on your vast experience with citrus fruits. Only in the plots of science fiction novels and Hollywood movies do we accept bizarre behavior from our accustomed surroundings. Chemists are no exception. They sort substances by a variety of distinguishing traits based on their observations. They then use those traits to investigate new substances. In this experiment you will study the behavior of some chemicals whose identity you know. You will then use what you learn to study and identify a substance whose identity is unknown (at least by you!). This is a classic approach in chemistry and one you will encounter again and again, because it is such a useful way to approach experimentation in the chemistry laboratory.

As you make your way through science courses, you will be continuously called upon to collect, organize, and analyze data. It is important that you develop data collection and organizing techniques that make your information easy to understand and refer to after you have completed an experiment. Laboratory write-ups usually require that you revisit what you have done, and it helps to have documented your activities in a coherent way.

Read through each experiment before you enter the laboratory and try to visualize what you will be doing, what information you will have to record, how you can tabulate data, and how you will analyze the data. Organizing the procedure into a flow chart will help you stay organized. Remember that it's better to record too much information than too little. Don't rely on your memory to help you recount what you did and why you did it.

At the beginning of your course, data collection tables will be provided for you. You will simply need to fill in the data as you collect it. Gradually, you will be expected to construct your own data collection tables. Pay attention to how the provided tables are organized so that you can profit from their example. It is convenient to organize data in such a way that you can compare information on a single page rather than flipping through several pages of your lab notebook looking for information.

The data you will collect in this experiment are called *qualitative* because they do not use numbers from measurements or calculations. In other experiments, you will collect numerical, or *quantitative*, data. The ways in which qualitative and quantitative data are organized depend on the type of experiment you are doing, the types of data you are recording, the relationships for which you are looking, and the analysis you plan to do. In this experiment, you will describe various physical properties of identified and unidentified solids and liquids. You will also examine their chemical reactivity toward each other and use both sets of properties to identify a solid of unknown identity.

Physical properties can be measured or observed without the substance reacting chemically. Physical properties include characteristics such as color, density, state (solid, liquid, or gas), solubility in liquids, melting point, boiling point, mass, volume, and size. *Chemical properties* are those properties the substance exhibits when it reacts with other substances or when its chemical composition is changed. Reactivity with acids and burning in oxygen are examples of chemical properties. You cannot simply look at the physical properties of a substance and make a statement about the chemical properties. Chemical properties are observed only when the substance is in the process of changing its composition—a chemical change. Physical changes can be carried out with no effect on the composition of the substance. Melting ice or dissolving sugar changes the physical state of the substance but not its chemical composition.

How can you tell whether a change is chemical or physical? Sometimes it is obvious: burning wood changes the wood into ashes and smoke. Sometimes it is not so obvious: dissolving sugar seems to "disappear," but has it changed chemically? What happens when the water it dissolved in evaporates? Is the sugar recovered intact? Usually, chemical reactions can be identified by the presence of one or more of the following: color change, production of heat, production of gas (fizzing), cooling of substance, production of light, the formation of a precipitate (two liquids are mixed and a solid is produced), and the production of or change in odor.

In this experiment, you will discover the chemical properties of a variety of solids of known identity by mixing them with liquids and examining their reactivity with each other. You will also describe the physical characteristics of each solid and liquid and look for changes in these when substances are brought together. You will then determine the identity of an unknown solid by referring back to the data recorded for known substances.

PRE-LAB EXERCISES

1. Generate a list of three real-life processes that are physical changes. Explain your choices.
2. Generate a list of three real-life processes that are chemical changes. Explain your choices.
3. Can you think of a real-life process that fits into both (or neither) categories? Explain.
4. Why should you never sniff things directly in the chemistry laboratory?
5. Look at the data table associated with part 2, liquid + liquid. Why are some of the boxes blacked out?

PROCEDURE

1. Physical Properties

On your data sheet, describe each solid and each liquid in terms of its physical characteristics. Examine each solid and liquid with a hand lens. Make sure to note the presence and shapes of any solid crystals. Note textures and colors. Liquids may differ in their consistency; note that. Note the presence of an odor; *however, don't sniff any substance directly.* Instead, wave the air above the substance toward your nose.

2. Chemical Properties

Solids + Liquids   On your data sheet, you will see a data table in the form of a matrix that lists solids in one direction and liquids in the other. You will use a well plate or test tubes to examine the chemical properties of each solid by mixing it with each of the liquids. Organize the materials in your well plate or test tubes in the same manner that the data table is organized. Make sure to label each well or test tube.

Use a spatula tip to put a very small amount of each solid in a well or test tube. Add a few drops of liquid. You may have to stir the substances with a stirring rod or toothpick. On your data table, record any changes you observe: color changes, gas formation, dissolution, and so on. Remember that the disappearance of a substance when mixed with a liquid does not necessarily indicate a chemical reaction; it may be that the solid is able to dissolve in the liquid, which is a physical property of the solid. If no reaction or physical change appears to have occurred, write NR for "no reaction."

After you have mixed all the solids with all the liquids and recorded your results, go back and add a few drops of the red cabbage extract to each well or test tube (except the ones that already have red cabbage extract present). Record the results in your data table.

Liquids + Liquids   On your data sheet, you will see a second data table that will allow you to tabulate the results of mixing the liquids with liquids. Again, you should organize the liquids in your well plate or test tubes in the same manner that the data table is organized. Make sure to label each well or test tube.

Mix a few drops of one liquid with a few drops of another. Mix with a toothpick or a stirring rod. Record your observations in the data table.

After you have mixed all the liquids with each other and recorded your results, go back and add a few drops of the red cabbage extract to each well plate or test tube (except the ones that already have red cabbage extract present). Record the results in your data table.

3. Identification of an Unknown Solid

Your instructor will provide you with an unknown solid. Record the number of the unknown on your data sheet. Briefly describe the physical appearance of your unknown. Use the data table provided to organize your well plate or test-tube rack. Mix a small amount of solid with each of the liquids. Record your observations. After mixing and recording your results, add a few drops of the red cabbage extract to each liquid mixture. Record your results. Compare the results you obtained with the results you obtained above. Can you identify your solid?

Data Sheet for Unit 2

Name_____

Date_____

1. Physical Properties

Solids

  a. Sodium chloride _____

_____

  b. Sodium bicarbonate _____

_____

  c. Calcium bicarbonate _____

_____

  d. Citric acid _____

_____

Liquids

  a. Distilled water _____

_____

  b. Vinegar _____

_____

  c. Ammonia _____

_____

  d. Mineral oil _____

_____

  e. Red cabbage extract _____

_____

2. Chemical Properties

Solids + Liquids

Solids → Liquids ↓	Sodium chloride (NaCl)	Sodium bicarbonate (NaHCO₃)	Calcium carbonate (CaCO₃)	Citric acid (C₆H₈O₇)
Distilled water				
+ Red cabbage extract				
Vinegar				
+ Red cabbage extract				
Ammonia				
+ Red cabbage extract				
Mineral oil				
+ Red cabbage extract				
Red cabbage extract				

Liquid + Liquid

Liquids →	Distilled water	Vinegar	Ammonia	Mineral oil	Red cabbage extract
Distilled water					
+ Red cabbage extract					
Vinegar					
+ Red cabbage extract					
Ammonia					
+ Red cabbage extract					
Mineral oil					
+ Red cabbage extract					
Red cabbage extract					

3. Identification of an Unknown Solid

Unknown number_____

Physical properties: _____

_____

_____

_____

Chemical Analysis

	Distilled water	Vinegar	Ammonia	Mineral oil	Red cabbage extract
Unknown solid					
+ Red cabbage extract					

Probable identity of unknown solid _____

Briefly explain your reasoning. _____

_____

_____

_____

_____

_____

_____

3

An Introduction to Volume and Mass Measurements

OBJECTIVES

- Become familiar with units of volume and mass.
- Learn to identify and use different types of laboratory volumetric ware.
- Make mass measurements using equipment available in your lab.
- Learn to differentiate among different calibration scales.
- Determine the number of significant decimal figures in a measurement.
- Learn how to report data in tables and graphs.

EQUIPMENT/MATERIALS

100-mL beaker, 100-mL volumetric flask, 250-mL Erlenmeyer flask, 100-mL graduated cylinder, 50-mL buret, index card with blackened area, laboratory balance, assorted pennies minted in different years.

INTRODUCTION

Let's start with some definitions:

- *Mass* is defined as a measure of the quantity of matter in a body.
- *Volume* is defined as the amount of space a body occupies in three dimensions.

You are probably more familiar with units of volume than with units of mass. In cooking, you typically make volumetric measurements in cups, teaspoons, and quarts. In the United States, we still buy our gasoline in gallons. Storage space is described in cubic feet. You are also familiar with metric volumetric measurements: you buy soda pop in 2-liter bottles, and the volume of automobile engines is described in liters as well. Medical personnel commonly refer to doses and other volumes in "cc's" of liquid; cc is the abbreviation for cubic centimeter, which is the same thing as a milliliter, mL.

Mass is a measure of the amount of material in an object; it is not the same thing as weight. *Weight* is a force and depends on gravity; *mass* is independent of gravity. Thus you would have the same mass on Jupiter as on the moon, but your weight would be considerably different. Similarly, if you were lucky enough to fly on the space shuttle, you might be weightless but you are never massless as long as you contain matter. Because all of the measurements you will make in this course will occur here on Earth, we will not distinguish further between mass and weight. Some of the most common measurements made in a chemistry laboratory are those of mass (milligrams, grams, and kilograms) and volume (milliliters and liters).

Measuring Volume

Measuring the volume of a liquid is easy, right? Everyone has poured liquid into some kind of measurement device at home—a spoon, a measuring cup, and so on. You already know that not all measuring devices and measurements are created equal. If a recipe calls for two cups of flour, you would not measure it out with a teaspoon, nor would you use a 2-liter bottle to measure out a dose of cough syrup. Similarly, if a recipe calls for ¼ teaspoon of cayenne pepper, you should probably use a measuring spoon that holds exactly ¼ teaspoon, rather than using a tablespoon and trying to guess how much a quarter teaspoon is.

The same issues of precision and appropriate measuring devices arise in a chemistry lab when you measure the volume of a liquid. Chemists measure liquid volumes with a variety of devices collectively called *volumetric ware*. Measuring volume in a laboratory requires using the proper piece of volumetric ware with the appropriate level of precision. In this unit you will examine the types, uses, and accuracies of a variety of common glassware. Through your examination you will learn their limitations.

Volumes in chemistry are most commonly measured in metric units of liters (L) or milliliters (mL). In some cases, usually clinical settings, measured volumes may be reported in units of deciliters or centiliters. Some health care professionals still use the cubic centimeter (cc or cm^3). A cubic centimeter is essentially the same thing as a milliliter.

Measuring Mass

Mass can be measured in the laboratory by using one of a variety of balances. Masses are most commonly reported in the metric units kilogram (kg), gram (g), and milligram (mg). The sample to be weighed is held in some container so that the mass recorded by the balance contains both the sample and the container. This container might be a beaker, a piece of weighing paper, or a flask. One typically measures the mass of the empty container on the balance, records it, and then records the mass of the sample plus the container. The *tared* mass of the empty container is then subtracted from the total mass of the sample and the container to obtain the net mass of the sample. Your notebook entry might look like the example in Figure 3.1.

There are a variety of balances in use in educational laboratories. Some report sample masses on a digital display, whereas others require the student to interpret a scale or series of calibration readings. Your instructor will provide specific details regarding the balances you will be using in your laboratory.

A few general guidelines can be established for balance use:

- Never use force to move any part of a balance. Doors, levers, pans, knobs, and so on, are all designed to move smoothly; if they don't, ask your instructor for assistance.

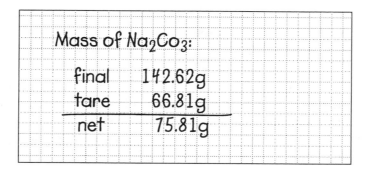

Mass of Na$_2$Co$_3$:

final   142.62g
tare   66.81g
net   75.81g

Figure 3.1   Sample notebook entry.

- Samples should be contained in appropriate holders unless you are instructed otherwise. Be sure that powder and liquid samples don't spill all over the balance as they could find their way into sensitive areas and cause damage.

- Use holders such as napkins, tongs, or tweezers when handling objects to be weighed. The oils from your skin can affect weights.

- Allow balances to stabilize before you take a reading. Swaying motions of balance pans and fluctuations of digital readings must stop before a measurement is made. Drafts and vibrations caused by student activity can cause balance readings to fluctuate; be aware of these influences.

- When you complete a measurement, return balance weights or knobs back to zero.

- When doing a series of measurements, always use the same balance in order to minimize equipment inconsistency. For example, weigh a beaker with a sample in it on the same balance that you used to weigh the empty beaker.

Significant Figures

Every measurement contains *uncertainty*, or error that comes from the technique or the equipment used to make the measurement. The degree of calibration or precision indicates how much uncertainty is associated with a measurement made using the device. For example, compare the uncertainty in the timing of a race using the two time-measuring devices shown in Figure 3.2. You would have a great deal of confidence in the time obtained using

Figure 3.2   Timing devices.

the digital stopwatch. However, the egg timer can only tell you whether the race took more than 3 minutes or less than 3 minutes.

The uncertainty in a measurement is always found in the last reported number. For example, a reported volume of 1.05 mL has uncertainty in the 5; the actual volume may be a little more than 1.05 mL or a little less than 1.05 mL. This uncertainty is indicated by reporting the number as 1.05 ± 0.01 mL. This tells you that the last number is a "best guess" made by the person doing the measurement, and that the actual volume is likely to be between 1.04 and 1.06 mL. By contrast, if the measurement was reported as 1.05 ± 0.05 mL, we would know that there was greater uncertainty associated with the measurement and the actual value could be anywhere between 1.00 and 1.10 mL.

The numbers included in a measurement are called *significant figures*. Do not confuse significant figures with decimal places. A measurement may contain 4 significant figures but have three decimal places—for example, 1.234. Conversely, a number might have four decimal places and only 2 significant figures: 0.0059. The measuring device you use and the amount contained determine both the number of significant figures and the number of decimal places you report in a measurement. Each device you use is designed to allow you to report a specific number of decimal places—use every one!

When students encounter measurements that contain zeros, they often have trouble determining which are significant and which are mere placeholders. Here are some guidelines regarding the significance of zeros:

- Zeros at the beginning of a number are decimal placeholders and are not significant.
- Zeros within a number are significant.
- Zeros at the end of a number with a decimal are significant.
- Zeros at the end of a number without a decimal are ambiguous; they may or may not be significant.

For example, there are 5 significant figures in the measurement below.

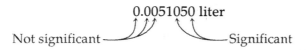

0.0051050 liter

Not significant ——— Significant

The zero dilemma can often be solved by using exponential notation. Thus the number above becomes 5.1050×10^{-3} L, and it's easy to see that it has 5 significant figures.

When performing calculations using measurements, remember that the result of an addition or subtraction has the same number of decimal places as the measurement with the fewest number of decimal places. For example:

11.48 g		44.977 mL
+ 0.0429 g		− 24.1 mL
11.52 g		20.9 mL

When multiplying or dividing, the measurement with the fewest number of significant figures determines the number of significant figures in the calculated result. For example, if a measurement with 4 significant figures is multiplied by a measurement with 3 significant figures, the result will be reported using only 3 significant figures. If you drive exactly

65.2 km/h (3 significant figures) for exactly 1.1178 hours (5 significant figures), the distance you covered can be determined by multiplying the rate by the time:

$$65.2 \text{ km/h} \times 1.1178 \text{ h} = 72.9 \text{ km}$$

If you perform this operation on your calculator, you will get the answer 72.88056; however, your calculator doesn't know that the uncertainty in the rate measurement carries over into uncertainty in the distance determination. The value 72.9 km is the best we can do.

PRE-LAB QUESTIONS

1. How many significant figures are in each of the following measurements?

   (a) 4.877 g (b) 0.004 mL (c) 5600.044 mm

   (d) 0.0092 kg (e) 0.009 20 kg (f) 4510 g

2. Perform the following calculations and report your answers with the appropriate units and number of significant figures:

   (a) 0.066 mL + 3.44 mL (b) 15.366 g ÷ 10.5 mL

   (c) 72.1 g + 0.0055 mg (d) 72.1 g + 0.0055 kg

PROCEDURE

1. Measuring Volume

There are numerous ways to measure volume. The simplest way is to pour the substance into a container such as a measuring cup and read the level of the substance off the side. However, when a volume of liquid is poured into a container, the liquid surface curves, as shown in Figure 3.3.

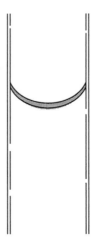

Figure 3.3   The meniscus.

The meniscus results because attraction between the liquid molecules and the walls of the container distorts the liquid line. When reading a volume, chemists take the measurement at the bottom of the meniscus. Because it is sometimes hard to distinguish the bottom of the meniscus, it can be made to stand out by placing a piece of black paper or an index card with a blackened area behind the glass below the liquid line. This outlines the meniscus very much as black eyeliner outlines an eye. The same effect can be accomplished with a small piece of black flexible tube that sits around the glass and can be moved easily up and down the column.

Several types of containers allow you to measure volume in a laboratory. Figure 3.4 shows a *beaker*, a *graduated cylinder*, an *Erlenmeyer flask*, a *volumetric flask*, and a *buret*—all of which can be used to measure volume.

Obtain one of each of the following and take it to your bench:

     100-mL beaker

     100-mL volumetric flask

     250-mL Erlenmeyer flask

     100-mL graduated cylinder

     50-mL buret

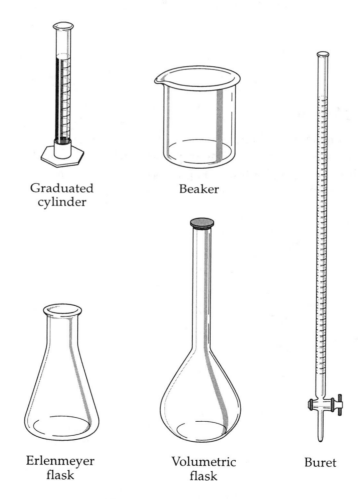

Graduated cylinder   Beaker

Erlenmeyer flask   Volumetric flask   Buret

Figure 3.4   Common pieces of glassware used to measure or contain liquids.

Carefully examine each of the five pieces of glassware used to measure volume, and answer questions 1–5 on your data sheet.

Each piece of glassware is calibrated differently. You cannot assume that the same types of glassware have the same calibrations. Always look at the scale and determine the value of the increments. Examine the calibrations on these five pieces of volumetric ware and answer question 6 on your data sheet.

Place 50 mL of water in the 100-mL beaker. Transfer it to the graduated cylinder. Is it exactly 50 mL? Record the more precise volume of the sample of water as determined from the graduated cylinder. As mentioned above, when determining how much volume is in a container, you need to read the bottom of the meniscus.

Be sure you take readings at eye level. Taking readings above or below eye level will give you different measurement values. Take three readings from the graduated cylinder at different levels, as shown in Figure 3.5. Record your results. How do the values differ?

Place 50 mL of water in the 250-mL Erlenmeyer flask. Transfer it to the graduated cylinder. Record the volume of the sample of water as determined from the graduated cylinder on your data sheet. Was the Erlenmeyer flask more useful than the beaker for measuring exactly 50 mL?

Measure out 100 mL of water in the graduated cylinder. Transfer it to the volumetric flask. Was this a closer match than you obtained with the transfer from either the beaker or the Erlenmeyer flask in the previous steps? Record your observations on your data sheet.

Your instructor will show you how to clamp your buret. Make sure that it is perfectly vertical (why?). When filling a buret, make sure that the stopcock is in a horizontal position or you may find yourself with a very wet bench top. Fill your buret with water through a funnel in the top opening. Be careful not to fill it too quickly because the liquid may back up and spill out over the top. When full, allow a small amount of liquid to drain from and fill the tip. Make sure there are no air bubbles in the tip. Use the buret to add 25.0 mL of water to the graduated cylinder. Do the two volumes correlate?

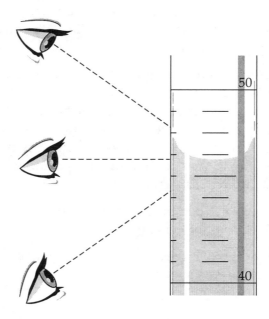

Figure 3.5   Liquid volume readings taken from different angles.

2. Measuring Mass

Your instructor will give you a collection of pennies from a variety of years. Sort them into the years they were minted. Record the year minted for each sample and the number of pennies for that particular year.

Weigh each batch of pennies that were minted in the same year. Try to avoid touching the pennies with your fingers. Record the masses for each batch. Note the following information about your balance: type, number of decimal places reported, and the procedure for reading measurements. (For example, "I step on the scale to get my weight, pick up the pig and get the combined weight, then subtract my weight from the combined weight to get the weight of the pig.")

From the data above, calculate the average mass of a penny for each year in your sample. Follow the sample calculation below.

$$\text{Average mass of 1 penny} = \frac{\text{Total mass of sample}}{\text{Number of pennies in sample}}$$

Mass of sample: 30.0567 g
Number of pennies in sample: 10 pennies

$$\frac{30.0567 \text{ g}}{10 \text{ pennies}} = 3.0057 \text{ g / penny}$$

For one group of pennies, weigh each individual penny and record each mass. Record the year the pennies were minted. Answer the questions about your measurements on your data sheet.

3. Preparation and Analysis of a Graph

Sometimes the relationships between two experimental quantities are not apparent until you organize and present them in a graph. The visual representation of data helps you to analyze your experiments and draw conclusions. The data you collected in the exercise above can be effectively presented in graphical form.

You will construct a commonly used graph called a "bar graph." Bar graphs consist of a series of vertical or horizontal bars, each of which represents a quantity or a number. Two examples of bar graphs are shown in Figure 3.6. The height of the vertical bar or the length of the horizontal bar indicates the value of the quantity shown on the scale that is parallel to it. The bar begins at zero and ends at the number or quantity it represents. The other scale identifies the bar—in our case, year of mint for penny samples. Bars can be colored, shaded, or divided to highlight different aspects of the quantity.

Your instructor may offer you a computer graphing option. If this is the case, he or she will provide appropriate directions for using the program.

Construct a bar graph for the average mass of each sample of pennies. Use your graph to answer the questions on the data sheet.

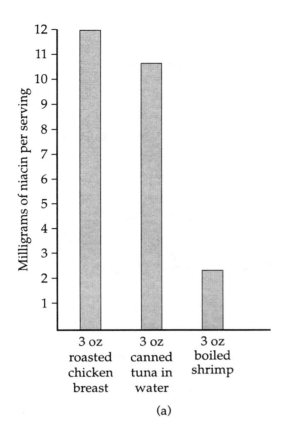

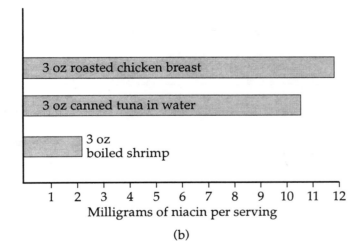

Figure 3.6 (a) Vertical bar graph with identifying labels on the horizontal axis.
(b) Horizontal bar graph with identifying labels on the bars.

Data Sheet for Unit 3

Name_____

Date_____

1. Measuring Volume

1. In your opinion, which piece of glassware will have the most uncertainty associated with it? _____

2. Which holds the most volume?_____

3. Which would you use to measure out 14.5 mL of water?_____

4. Which would you use to measure out ~40 mL[1] of water?_____

5. Which would you use to make exactly 100.0 mL of a solution?_____

6. Fill in the table below in which you answer each of the following questions for each piece of glassware.

   a. What values do the incremental marks for each piece of glassware represent?

   b. How many decimal places can be reported for each piece of glassware?

   c. How many significant figures can be reported for each type of glassware when it contains 20 mL of water?

Glassware	Incremental marks	Number of decimal places	Significant figures for 20 mL of water
100-mL beaker			
100-mL volumetric flask			
250-mL Erlenmeyer flask			
100-mL graduated cylinder			
50-mL buret			

50 mL of water measured in the beaker actually corresponds to _____ mL when measured with the graduated cylinder.

Meniscus read from above_____

Meniscus read at eye level_____

Meniscus read from below_____

How do these values differ?

50 mL of water measured in the 250-mL Erlenmeyer flask actually corresponds to _____ mL when measured with the graduated cylinder.

Which is more useful for measuring out exactly 50 mL, the beaker or the Erlenmeyer?

_____

[1] The symbol ~ is used to mean *approximately* in most scientific settings. You might also see the expression *ca. 20 mL* which means the same thing. *Ca.* is an abbreviation for the Latin word *circa*, which means *around*.

How does 100 mL of water measured in the 100-mL graduated cylinder match to the value of the volumetric flask? _____

How does 25 mL of water dispensed by the buret match to the value of the graduated cylinder? _____

2. Measuring Mass

Weighing Pennies

Year of penny	Number of pennies	Total mass	Average mass

Balance type_____

Number of decimal places reported_____

How do you use the balance for making measurements? _____

_____

Individual masses for the year_____

Sample	Mass
1	
2	
3	
4	
5	
6	

Average Mass _____

Are there variations in the average mass of your pennies? _____

For the sample in which you recorded the individual penny masses, how does the average calculated mass of each penny compare to the measured mass? _____

_____

Are the masses constant from year to year? _____

Are the individual masses constant within a sample? _____

3. Preparation and Analysis of a Graph

1. What relationship does your graph show between the year a penny was minted and its average weight?

2. Is the mass of a penny constant over the span of time you analyzed?

3. What factors could contribute to the results shown in your graph?

4. List three factors that could affect the mass of a penny, and suggest ways to determine their influence.

5. The United States Mint is the agency responsible for setting the metal composition of coins. Visit their Web site at http://www.usmint.gov/, and link to About the Mint → Fun Facts → The Composition of the Cent to access information about Lincoln pennies. How does this information help to explain your experimental results?

4

A Lesson in Density

OBJECTIVES

- Learn about density and conversion factors.
- Determine the density of a liquid (ethanol) and of a solid (pennies).
- Begin to learn how to organize data in a table.
- Learn how to construct a line graph.

Review

- Use of volumetric ware and volume measurements.
- Use of weighing equipment and mass measurements.
- Calculating averages.

EQUIPMENT/MATERIALS

50-mL burets, 50-mL graduated cylinder, 25-mL graduated cylinder, 10-mL pipets, 10-mL graduated cylinders, weighing boat, pennies from various years, distilled water.

INTRODUCTION

In the previous experiment, you learned how to measure mass and volume. Mass and volume are properties that change depending on the size of a sample; they are *extensive properties*. Extensive properties are not useful in classifying a substance. Two samples of the same substance, say iron, can have different masses, just as a 1-L bottle of cola can taste very different from a 1-L bottle of orange soda, even though they have the same volume. By contrast, *intensive properties* are inherent in a sample and *are* useful in identifying a substance. A substance will exhibit the same intensive properties no matter what the sample size. Even though mass and volume are extensive properties, the ratio of the mass of a substance to its volume defines the intensive property *density*.

Remember the old joke, "Which is heavier, a pound of lead or a pound of feathers?" Even though a pound is always a pound, we *think* of lead as being heavier than feathers because it has such a high density. The density of a substance is defined as the mass per unit volume. In chemistry, we define density more specifically as the mass of 1 mL or 1 cubic centimeter (cc or cm^3) of the substance. Mathematically, density is defined as follows:

$$\text{Density (g / mL)} = \frac{\text{Mass (g)}}{\text{Volume (mL)}}$$

If there is a large mass in a small volume, the value of the density is large. From a scientific standpoint, this means that the atoms or molecules have a large mass and are packed closely together with not much space between them. Metals fall into this category. The atoms of metals have lots of protons and neutrons and thus have a large mass (e.g., lead atoms have around 207 neutrons and protons!). These atoms are also strongly attracted to each other so they are nestled closely together. Thus, there is a large amount of matter (mass) in a small volume: high density. Conversely, if there is a small mass spread out over a large volume, the density is low. This means that the molecules are relatively light or not packed very tightly together. Helium is an atom of very low mass (only 2 protons and 2 neutrons), and because it is a gas at most temperatures, there is a lot of space between the atoms. Thus helium has a very low density and can float above air.

Whether an object floats or sinks relative to another is directly related to its density—*not* its mass. A 4.0-g cork will float on water because it is less dense than water. Suppose the cork has a mass of 1000 kg. This very large cork would still float because its density is still less than that of water. Conversely, even a tiny, 0.005-mg piece of iron will sink in water because it has a higher density. The fact that density is independent of the amount of the substance present is exactly why it is an intensive property and can be used to describe a substance. Some liquids have a greater density than others. Can you think of a liquid that is less dense than water? More dense than water?

Density is useful for a variety of purposes. It can be used to determine relative amounts of substances dissolved in aqueous solutions (the basis for the specific gravity measurements of urine that are used in clinics); it can be used to identify a substance (you will do this in this laboratory unit); and it can be used as a conversion factor to move between mass and volume, as illustrated below.

Mass (g) = Density (g/mL) × Volume (mL)

Volume (mL) = Mass (g) ÷ Density (g/mL)

Determining the density of a liquid or solid sample is relatively easy, but the technique is different for each. In this laboratory unit, you will measure the density of a pure liquid, solutions of sugar in water at different concentrations, and a sample of pennies. You will also graph the relationship between the year a penny was minted and its density.

PRE-LAB QUESTIONS

1. You measure out 45.0 mL of a dense liquid. Its mass is 63.3 g. What is the density of this liquid?
2. A density column is a collection of different liquids and/or solids of different densities in a glass column. It is useful in exploring qualitative differences in density by observing

what floats on what. A density column in a graduated cylinder is made from the following substances/items. Draw a rough picture of what this density column would look like.

Substance/Item	Density (g/mL)
Water (*l*)	1.000
Ether (*l*) (diethyl ether)	0.715
Carbon tetrachloride (*l*)	1.594
Iron nail	7.8
Cork	0.44

PROCEDURE

1. Measuring Masses and Volumes of Ethanol Samples

To determine the density of a liquid sample, you need to measure both the volume of the liquid and its mass.

Tare a 10-mL or a 25-mL graduated cylinder on your balance. If your balance cannot accommodate the height of a graduated cylinder, tare a 25-mL beaker and transfer your liquid back and forth between the beaker (to measure the mass) and the graduated cylinder (to measure the volume).

Measure out a specific volume of ethanol, around 2 mL. You do not need to measure out exactly 2 mL. It is more efficient to add approximately 2 mL of ethanol to the graduated cylinder and then record the exact volume on your data sheet. Weigh the graduated cylinder containing the ethanol and determine the mass of the ethanol.

Repeat this process with ~4 mL, ~6 mL, and ~8 mL of ethanol.

After you have collected the data from your measurements, calculate the density of each of your samples. Use the example below as a guide.

Example Calculation

Mass of sample: 7.00 g
Volume of sample: 8.88 mL

$$\text{Density (g / mL)} = \frac{\text{Mass (g)}}{\text{Volume (mL)}} = \frac{7.00 \text{ g}}{8.88 \text{ mL}} = 0.788 \text{ g / mL}$$

Calculate the average density of ethanol. Record the value.

2. Solution of Sugar and Water

Weigh out 20.0 g of sugar.

Weigh a small (100-mL) beaker and record its mass. Weigh out 20.0 g of water, determine its volume in milliliters, and calculate its density. Record this information.

Add the sugar to the water and stir gently until it is *completely* dissolved (this may take a few minutes, so be patient). Determine the final mass and volume of the solution. Calculate the density of the solution. (Take care to factor out the mass of the container, which you measured above.)

Determine the mass of 10.0 mL of your sugar–water solution.

Determine the mass of 15.0 mL of your sugar–water solution.

Determine the mass of 20.0 mL of your sugar–water solution.

Determine the mass of 35.0 mL of your sugar–water solution.

Repeat this entire process using 25.0 g of sugar and 15.0 g of water.

3. Determining the Density of a Penny

Solids do not take on the shape of their container, so it is a bit trickier to determine the volume of a solid unless it has a very regular shape. One way to obtain the volume occupied by a solid sample is to use the displacement method (just like Archimedes—eureka!). The sample in question is weighed and the mass is recorded. A volume of water is measured into a container such as a graduated cylinder. This volume must be enough to cover the solid when it is immersed. The initial volume is recorded, and the solid is added to the water. Adding the solid causes the water level to rise as the solid displaces water. The second volume is read, and the difference between the first and second volumes is the volume occupied by the solid. (See Figure 4.1.)

The density can be calculated by dividing the mass of the sample by the volume change in the water. Use the example at the top of the next page as a guide.

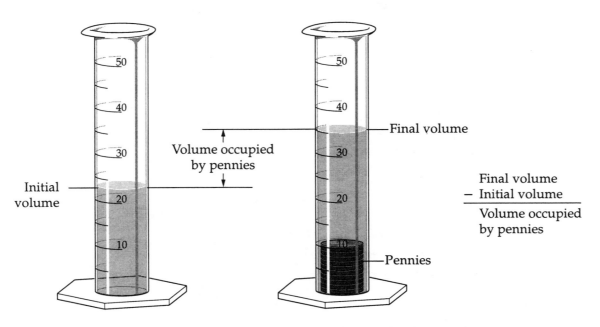

Figure 4.1   Measuring the volume of a solid by displacement.

Example Calculation

Mass of sample: 85.76 g   Final volume: 21.56 mL
                                 Initial volume: $-$10.01 mL
                                 Volume change: 11.55 mL

$$\text{Density} = \frac{\text{Mass (g)}}{\text{Volume change (mL)}} = \frac{85.76 \text{ g}}{11.55 \text{ mL}} = 7.43 \text{ g / mL}$$

Your instructor will give you a sample of pennies from two different years. You will combine your results with those of other class members to determine if there are any density trends over several decades.

Use different numbers of pennies, and generate four sets of mass and volume measurements for each year. Make sure that the number of pennies in a sample produces enough of a volume change so that you can record a measurement.

Weigh each sample, and measure the volume by the displacement method outlined above. You will need to use a graduated cylinder that is large enough to accommodate your sample.

After putting the pennies into the water, tap the bottom of the graduated cylinder gently against the bench to free any trapped air bubbles.

Enter your data on the data table.

Calculate the average density for each sample of pennies. Record this value in the class data table.

4. Graphing the Relationship Between Density and the Year a Penny Was Minted

In the previous experiment, you examined the mass changes in pennies over several decades. Factors other than changes in the chemical composition of a material can affect mass measurements. Density, because it is an intensive property, is a reliable indicator of compositional changes.

Your instructor will outline the steps involved in constructing a line graph. Use the combined data from the whole class to construct a simple line graph. Use your graph to answer the questions on your data sheet.

Unit 4 Data Sheet

Name_____

Date_____

1. Measuring Masses and Volumes of Ethanol Samples

Mass of tared graduated cylinder _____

Sample	Exact volume	Total mass	Net mass*	Density**
Graduated cylinder	N/A		N/A	N/A
~2 mL				
~4 mL				
~6 mL				
~8 mL				
			Average density	

*Obtained by subtracting mass of graduated cylinder.

**Obtained through calculation $d = m/V$.

Answer the following questions:

1. Was there variation in the densities of your individual samples?

_____

_____

2. How does changing the volume of a sample affect the mass of the sample?

_____

_____

3. How does changing the volume of a sample affect the density of the sample?

_____

_____

4. Ethanol evaporates readily at room temperature. How does the evaporation affect the mass? How does it affect the volume? How does it affect the density?

_____

_____

2. Solution of Sugar and Water

First solution of sugar and water (20 g sugar + 20 g water)

Mass of beaker _____

Volume of 20.0 g of water _____

Density of 20.0 g of water _____

Volume of 1st sugar–water solution _____

Mass of 1st sugar–water solution _____

Density of total 1st sugar–water solution _____

Volume of 1st sugar-water solution	Mass of sugar–water solution	Density of sugar–water solution
10.0 mL		
15.0 mL		
20.0 mL		
35.0 mL		

Second solution of sugar and water (25 g sugar + 15 g water)

Mass of beaker _____

Volume of 15.0 g of water _____

Density of 15.0 g of water _____

Volume of 2nd sugar–water solution _____

Mass of 2nd sugar–water solution _____

Density of total 2nd sugar–water solution _____

Volume of 2nd sugar-water solution	Mass of sugar–water solution	Density of sugar–water solution
10.0 mL		
15.0 mL		
20.0 mL		
35.0 mL		

Answer the following questions:

1. When you mix two substances together, are the volumes additive? Use your experimental evidence to support your answer.

_____

_____

_____

2. When you mix two substances together, are the masses additive? Use your experimental evidence to support your answer.

_____

_____

_____

3. Use results from your experiment to demonstrate that density is an intrinsic property.

_____

_____

_____

4. Determining the Density of a Penny

Prepare a table with the headings below to record your data.

Year of penny sample   Mass (g)   Volume (mL)   Density (g/mL)

Determine the average density per year, and add it to similar data from the rest of the class.

Prepare a second table with the headings shown below to record the data from the entire class. Then construct a line graph, plotting the density as a function of year minted. Attach your tables and graph to your report.

Year of penny sample   Density (g/mL)

Use your graph to answer the following questions:

1. What trends are indicated by the graph?

_____

_____

_____

2. How does the density graph compare to the mass graph from the previous experiment?

_____

_____

_____

3. How do the trends in density compare to compositional changes in pennies?

_____

_____

_____

5

Ionic Reactions: Precipitates, Solubility, and Metal Activity

OBJECTIVES

- Carry out a series of ionic reactions to determine which ones form precipitates.
- Make statements about the solubilities of ionic compounds based on experimental results.
- Write net ionic equations for precipitation reactions.
- Predict formulas of ionic compounds.

EQUIPMENT/MATERIALS

Test tubes and test-tube rack (or well plate), solutions of ionic compounds, concentrated aqueous ammonia.

INTRODUCTION

Many things dissolve in water, which is why water is often called the "universal solvent." Solutions in which water is the solvent are called *aqueous solutions*. You drink aqueous solutions (coffee, tea, soda), cry an aqueous solution (tears), and secrete aqueous solutions (sweat, urine). Water is a particularly good solvent for ionic substances because of its high polarity. You are probably familiar with how easily NaCl (table salt), an ionic compound, dissolves in water. When ionic compounds dissolve in water, the ions separate. Each individual ion is surrounded by (and interacts with) water molecules, as shown in Figure 5.1. This separation of ions is known as *dissociation*. The dissociation of ions in water can be written using a chemical equation as follows:

$$NaCl(s) \longrightarrow Na^+(aq) + Cl^-(aq)$$

The symbols (*s*) and (*aq*) refer to the physical state of the substance and are known as *state symbols*. In this case, (*s*) refers to the solid state and (*aq*) refers to the aqueous state.

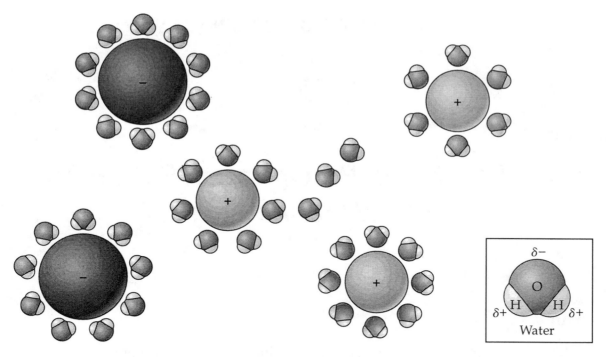

Figure 5.1   Dissolved ions surrounded by water molecules.

Rather than write the formulas of individual ions, the formula NaCl(*aq*) can represent the aqueous NaCl solution.

Not all ionic substances are soluble in water. If you place a piece of marble, which is composed mostly of calcium carbonate ($CaCO_3$), in pure water, little of it will dissolve. Indeed, rain would be catastrophic for marble buildings and statues if marble were readily soluble in water. What makes one ionic substance soluble and another not has to do with a complex balance of different energies. Sometimes the solution is a lower energy situation and the ions dissociate as they dissolve in the water. Other substances are in a lower energy state when their ions stay in the solid lattice. In this laboratory unit, you will explore solubility through *precipitation reactions*.

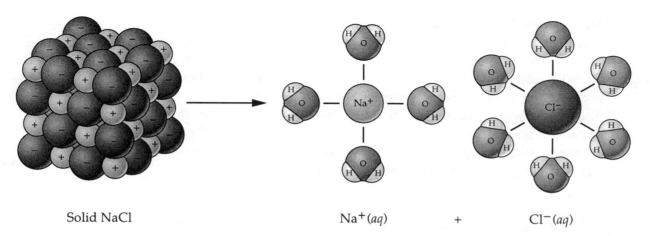

Solid NaCl   Na$^+$(*aq*)   +   Cl$^-$(*aq*)

Figure 5.2   Solid NaCl dissolves to form Na$^+$ and Cl$^-$, which are surrounded by water molecules (*aq*).

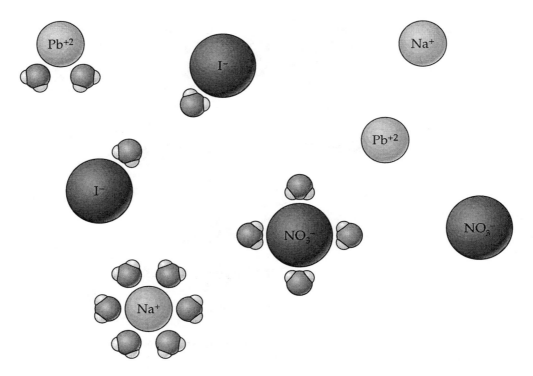

Figure 5.3   When $Pb(NO_3)_2$ and NaI solutions are mixed together, the resultant solution contains all ions.

When two aqueous solutions containing different ions are mixed together, the dissociated ions are able to interact with each other. For example, a lead nitrate solution contains dissociated Pb^{2+} and NO_3^- ions, and a sodium iodide solution contains dissociated Na^+ and I^- ions. If these solutions are mixed together, Pb^{2+}, Na^+, I^-, and NO_3^- ions will all bump around together in the same solution, as shown in Figure 5.3. The presence of four different ions in an aqueous environment leads to the following two chemical possibilities: (1) all the ions are attracted to the surrounding water sufficiently to stay dissolved, or (2) oppositely charged ions are attracted to each other more strongly than they are to the surrounding water and combine to form an insoluble solid. The formation of a solid is characterized by the appearance of either a milky cloudiness or a colored substance that was not present in either of the initial solutions. An insoluble product formed in this way is called a *precipitate*, designated (ppt), and the reaction is called a precipitation reaction.

In the case of NaI and $Pb(NO_3)_2$, both solutions are initially clear, but when they are mixed, a beautiful, glittery-yellow solid appears due to the formation of a product, the precipitate (see Figure 5.4). The question is, which ions have combined to form the precipitate? There are three possibilities:

1. Pb^{2+} combined with I^-: Pb^{2+} + $2I^-$ $\rightarrow$ $PbI_2(s)$
   lead ion   iodide ion   lead iodide

2. Na^+ combined with NO_3^-: Na^+ + NO_3^- $\rightarrow$ $NaNO_3(s)$
   sodium ion   nitrate ion   sodium nitrate

3. Pb^{2+} combined with I^- *and* Na^+ combined with NO_3^-.

(We know that Pb^{2+} didn't combine with NO_3^- or Na^+ with I^-. How do we know this?)

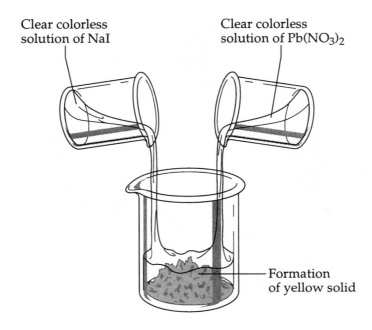

Clear colorless solution of NaI

Clear colorless solution of $Pb(NO_3)_2$

Formation of yellow solid

Figure 5.4   Formation of a precipitate.

If we were to examine a sample of sodium nitrate, we would find that it is a white solid that is very soluble in water. Conversely, lead iodide is a yellow solid that is virtually insoluble in water. From this information, we can deduce that PbI_2 is the yellow precipitate that forms. The balanced overall reaction describing the formation of this precipitate from the solutions described above can be written as follows:

$$Pb(NO_3)_2(aq) + 2\ NaI(aq) \longrightarrow PbI_2(s) + 2\ NaNO_3(aq)$$

Remembering that aqueous salts are dissociated into their constituent ions, we can rewrite this equation as:

$$Pb^{2+}(aq) + 2\ NO_3^-(aq) + 2\ Na^+(aq) + 2\ I^-(aq) \longrightarrow PbI_2(s) + 2\ NO_3^-(aq) + 2\ Na^+(aq)$$

We can cancel out the nitrate and sodium ions because they appear on both sides of the equation:

$$Pb^{2+}(aq) + \cancel{2\ NO_3^-(aq)} + \cancel{2\ Na^+(aq)} + 2\ I^-(aq) \longrightarrow PbI_2(s) + \cancel{2\ NO_3^-(aq)} + \cancel{2\ Na^+(aq)}$$

which leaves us with

$$Pb^{2+}(aq) + 2\ I^-(aq) \longrightarrow PbI_2(s)$$

This is called the *net ionic equation*. It shows only the reaction between the ions that form the insoluble product. In the solution, Na^+ and NO_3^- ions are present along with Pb^{2+} and I^- ions; however, they stay dissolved and do not participate in any chemical reaction. They are called *spectator ions* and are not shown in the net ionic equation.

In this laboratory unit you will mix various ionic solutions and fill in a data table to determine which ionic combinations produce precipitates. You will use your data to make some general statements about the solubilities of various ionic compounds. All the precipitates in this unit result from an exchange of partners between reactants. Let's look at a general scheme for exchange reactions:

$$M_aX_b + N_cY_d \longrightarrow \underline{\hspace{3cm}}$$

One possibility is that nothing happens. You mix the two solutions together and nothing changes. This indicates not only that M_aX_b and N_cY_d are soluble but that the compounds that M forms with Y and N forms with X are also soluble. For example, no precipitate forms when $NaCl(aq)$ and $KNO_3(aq)$ are mixed. The reaction describing this combination can be written as follows:

$$NaCl(aq) + KNO_3(aq) \longrightarrow NaNO_3(aq) + KCl(aq)$$

From this, you can conclude that NaCl, KCl, $NaNO_3$, and KNO_3 are all soluble in water. Thus any soluble ionic substance will not be observed as a precipitate.

The other possibility is that a precipitate does form. This means that either the compound that M forms with Y or the compound that N forms with X is insoluble. But which is it? This is where you must play chemical detective. You must find another circumstance to react M with Y or N with X to see if you get a precipitate. This will give you information about the original reaction.

Below are some guidelines to help you in this laboratory unit. These guidelines do not necessarily apply to other solutions outside of this unit.

Guidelines for Determining Ions That Form Precipitates in This Experiment

1. When (aq) is written after a compound's formula, this indicates that it is a water-soluble compound.

2. Once a compound is determined to be soluble in water, it will always be soluble in water.

3. Once a precipitate is identified as a water-insoluble compound, it will always be insoluble in water.

4. Any solution that is transparent contains a compound that is soluble.

5. When two solutions are mixed and you *do not* see a precipitate, then both of the new ionic combinations are soluble in water.

6. When two solutions are mixed and you *do* see a precipitate, one of the new combinations of the ions forms the precipitate.

Complex Ions

Sometimes metal ions will react with ammonia to form what is known as a *complex ion* or a *coordination complex*. A complex ion has a central metal ion with one or more molecules *coordinated* (bonded) to it. For example, nickel complexes readily with ammonia:

$$Ni^{2+}(aq) + 6\,NH_3(aq) \longrightarrow [Ni(NH_3)_6]^{2+}(aq)$$

The complex ion $[Ni(NH_3)_6]^{2+}$ is depicted in brackets because it is an integral unit. Complex ions are typically very soluble in water. Often they are so soluble that their formation will allow insoluble compounds to dissolve. For example, $NiCO_3$ is insoluble but addition of concentrated ammonia will cause it to dissolve because the complex ion will form:

$$NiCO_3(s) + 6\,NH_3(aq) \longrightarrow [Ni(NH_3)_6]^{2+}(aq) + CO_3^{2-}(aq)$$

In this experiment you will carefully add concentrated aqueous ammonia to some of your precipitates to see if the metal ions form a complex ion with ammonia.

PRE-LAB QUESTIONS

The following results were obtained by a student doing a similar experiment (using some metal ions that are environmentally hazardous!):

	NaI	NaSO₄	Ba(NO₃)₂	Pb(NO₃)₂	NiCl₂
NaI		NR	NR	Yellow ppt	NR
NaSO₄			Sol'n turned cloudy, white ppt.	Grayish-white solid	NR
Ba(NO₃)₂				NR	NR
Pb(NO₃)₂					Cloudy white

Based on her results, identify each of the following as soluble or insoluble in water:

a. BaI_2 _____

b. PbI_2 _____

c. NiI_2 _____

d. $BaSO_4$ _____

e. $NiSO_4$ _____

f. $PbSO_4$ _____

g. $NaNO_3$ _____

h. $NaCl$ _____

i. $NiNO_3$ _____

j. $BaCl_2$ _____

PROCEDURE

1. Testing Solubilities

Obtain a well plate or a rack of small test tubes. Arrange and label your test tubes in accordance with the matrix on the data sheet at the end of this experiment. Place a small amount of the solution corresponding to each row in the well plate or test tube (a few drops for a well plate or ~0.5 mL for a test tube). For each solution listed in the columns across the top, add a similar amount to the test tube or well plate and record your observations.

Before mixing and recording your results, make a note in the title boxes of any distinguishing characteristics of the test solutions (e.g., "colorless sol'n"). When you record your results of mixing the solutions, write NR in the boxes if there is no reaction. If there is a reaction, indicate briefly what you observe and indicate the possible identity of the precipitate, for example, "white cloudy ppt. AgOH or $NaNO_3$." In a chemical reaction, A + B will give the same results as B + A, so you need not repeat each test, nor do you need to react a substance with itself. Thus, half the matrix is unnecessary (which is why we made it gray).

After you have completed the tests, determine which substances are soluble and which are insoluble. If there is doubt or confusion, you may need to repeat part of the experiment. For this reason, it is imperative that you sort out the solubilities *before* you leave the laboratory.

2. Addition of NH₃

In each of the wells or test tubes where a precipitate has formed, *carefully* add a few drops of concentrated aqueous ammonia. Observe what happens and record your observations.

> **CAUTION: CONCENTRATED AQUEOUS AMMONIA CAN BE VERY CAUSTIC. USE IT IN A WELL-VENTILATED AREA. AVOID SKIN CONTACT. IF YOU DO GET IT ON YOUR SKIN, RINSE THOROUGHLY WITH LARGE AMOUNTS OF WATER.**

Data Sheet for Unit 5

Name_____

Date_____

1. Matrix for Analysis of Precipitation Reactions

	NaOH	NaNO$_3$	FeCl$_3$	AgNO$_3$	KCl	K$_2$CO$_3$	Cu(NO$_3$)$_2$
NaOH							
NaNO$_3$							
FeCl$_3$							
AgNO$_3$							
KCl							
K$_2$CO$_3$							
Cu(NO$_3$)$_2$							

Based on your results and a little deductive reasoning, list the ionic compounds that are and are not water-soluble:

Water-soluble   **Water-insoluble**

_____   _____

_____   _____

_____   _____

_____   _____

_____   _____

_____   _____

_____   _____

_____   _____

_____   _____

Look over your data to see if any ions are consistently present or absent from the precipitates. Can you make any generalizations about the solubilities of compounds containing the ions listed below?

a. Nitrate ion

b. Sodium ion

c. Potassium ion

d. Chloride ion

e. Silver ion

f. Hydroxide ion

g. Carbonate ion

2. Addition of NH₃

Probable identity of precipitate	Observations upon addition of NH₃
_____	_____
_____	_____
_____	_____
_____	_____
_____	_____

List the substances that form complex ions with NH₃. Check your list with your instructor and she or he will give you the formulas of these complex ions. Use the formulas to write equations for their formation.

6

Introduction to Experimentation: The Ice-Cube Dilemma

OBJECTIVES

- Explore the realm of scientific experimentation and the scientific method.
- Draw conclusions from macroscopic observations.

EQUIPMENT/MATERIALS

Ice cubes, colored ice cubes, sugar, other potential solutes, water, stopwatch or watch/clock with a second hand, regular soda (with sugar), diet soda, beakers, water.

INTRODUCTION

Are you a soda drinker? Have you ever noticed that ice melts faster in diet soda than in the sugared variety? Try it and see. The first stage of scientific experimentation is noticing something that is odd, unusual, or provokes a question: Why does ice melt faster in a diet soda? Scientists are often portrayed as clinical, logical, somewhat robotic beings. In fact, scientists are typically extremely curious—almost obnoxiously so. While most people might be mildly interested in why Ajax® cleanser turns blue when it gets wet or how a sunscreen protects you from getting sunburned, a chemist will find these questions fascinating and will not rest until he or she knows the answer. Often the answer can be found in the library or on the Internet. Asking one's chemist colleagues is another way to find answers. However, sometimes, the answer cannot be found in this manner because the answer is simply not known. This is when a scientist heads for the laboratory to do experiments.

Experimentation is not something that a good scientist will jump into without careful planning, even though jumping in can be extremely tempting! She should know what experimental results others have observed in that particular area of study, so that she doesn't simply repeat the work of others. It is also important to design experiments carefully. Experimental design is probably the most complex aspect of science. Among the factors that you have to consider when designing an experimental investigation are the following:

- Will I be able to obtain clear measurements or observations? Will I know what to look for?

- What variables[1] will I examine? How will I change these variables?

- Will I need a control experiment?[2] If so, how will I set that up for best comparison?

- What data will I examine? How will I accumulate those data? How will I analyze those data?

- Will the results of the experiments give me useful information? Will they answer the questions that I am asking? How will I be able to draw conclusions from my experimental results?

The results of one experiment will often lead to the planning of the next experiment. Suppose you are studying the effects of a group of different chemicals on mold growth in bread. Seeing that two of the substances you are examining both inhibit mold growth individually, you might want to try them in combination. An experiment that shows mold growth inhibition by a particular chemical might prompt you to vary the concentration of that chemical to see if there is an upper or lower limit where the effect is observed. If topical application of a chemical does not inhibit mold growth, will baking a chemical into the bread result in a different effect? Sometimes time and expense must come into play and some avenues of experimentation must be ignored in favor of others.

You can see that experimental design can be rather complex. On the other hand, it can be an extremely interesting and rewarding endeavor. Despite what you might think, *no experiment is ever a failure*. You will always learn something, even if it is just that you need to do a better job of designing experiments!

Let's get back to our ice dilemma: Why does ice melt faster in diet soda than in sugared soda? In this experiment, you will explore this question, first by verifying that it does indeed happen (no point in just taking our word for it!). Then you will do some simple experiments to determine whether it is a universal phenomenon and to try to figure out why it happens.

---

[1] Experimental *variables* are the aspects that you change from trial to trial. For example, if you were studying how fast a particular chemical reaction progresses, you might vary the concentration of the reagents, the temperature of the reaction, whether light and/or oxygen is present, and so on. Those factors are all variables, but be careful not to vary them all at once! (See the second note, below.)

[2] The purpose of a *control experiment* is to develop background information so that you can compare your experimental results to a situation where you have carefully "controlled" the variables. For example, if you want to see the effect of UV light on a particular species of fish, you will have to raise some of the fish in the absence of UV light. Moreover, you need to make sure that all other variables are the same for the two samples of fish. The temperature, pH, feeding schedule, oxygen concentration, and so on, have to be carefully controlled to ensure that any effects you observe can be traced to the UV light and not some other difference between the two populations.

PROCEDURE

1. Verification Experiment

First you will test the premise of this problem, comparing the ice-cube melting rate in regular versus diet soda.

Obtain two 250-mL beakers. Label one "diet" and the other "regular." Measure 100 mL of each of the sodas into its designated beaker. Make sure the sodas are the same temperature; they can be warm or cold, but they must be the same.

Select two ice cubes; make sure they are the same size. Place one ice cube in each beaker and begin timing. When each ice cube is completely melted, record the time on your data sheet.

2. Exploration Experiments

In this part, you will carry out a series of experiments to determine (a) whether the results you got with the diet and regular sodas above carry over into other circumstances, and (b) what is causing the ice to melt faster in one than in the other.

Does it work in all situations?

Measure 100 mL of water into one of the beakers. Fill a 100-mL beaker about a quarter full (~25 mL) of sugar. Add the sugar to the water and stir until completely dissolved. Label the beaker. Measure 100 mL of plain water into another labeled beaker. Let the two beakers sit on the lab bench for about 10 minutes so that you are sure they are the same temperature.

Select two ice cubes; make sure they are the same size. Place one ice cube in each cup and begin timing. When each ice cube is completely melted, record the time on your data sheet.

Repeat the process outlined above, substituting other solutions for the sugared water. Your instructor will provide assorted solutes and give you instructions for how many you should test. In each case, make sure your ice cubes are the same size and your two liquids are at the same starting temperature.

What is causing this to happen?

Repeat the process outlined above (with sugared water versus plain tap water). Use the colored ice cubes. Rather than timing the ice cube melting, observe the melting process carefully. What is different about how the ice cube melts in the plain water compared to how it melts in the water with sugar in it?

Repeat the process outlined above. Time the ice cube melting, but this time, *stir* the two samples thoroughly as the ice melts. Record your results.

What other experiments can you think of?

Consult with your instructor to design other experiments to try. If time and materials are available, perform your experiments and record your results.

Data Sheet for Unit 6

Name_____

Date_____

1. Verification Experiment

Time required for ice cube to melt in regular soda   _____

Time required for ice cube to melt in diet soda   _____

2. Exploration Experiment

Time required for ice cube to melt in ordinary water   _____

Time required for ice cube to melt in sugared water   _____

Time required for ice cube to melt in ordinary water   _____

Time required for ice cube to melt in _____   _____

Time required for ice cube to melt in ordinary water   _____

Time required for ice cube to melt in _____   _____

Time required for ice cube to melt in ordinary water   _____

Time required for ice cube to melt in _____   _____

Time required for ice cube to melt in ordinary water   _____

Time required for ice cube to melt in _____   _____

1. Does this process seem to be universal? Briefly explain.

   _____

   _____

   _____

2. Describe your observations for the melting of the colored ice cubes in plain tap water versus sugared water. Can you suggest an explanation for what you observe?

   _____

   _____

   _____

3. Report your results when you timed the melting ice while stirring the two samples.

_____

_____

_____

_____

_____

4. Postulate as to why ice melts faster in plain water than in water with sugar or salt dissolved in it. How does stirring change that?

_____

_____

_____

_____

_____

5. Why did you have to compare each sample to plain water each time? Why couldn't you just use the time from the first experiment?

_____

_____

_____

_____

_____

6. How would your results have been affected had you not paid attention to the initial temperature of the water?

_____

_____

_____

_____

_____

7. What other experiments did you design? What were they expected to show? Report on your results and any conclusions you can draw from your results.

_____

_____

_____

_____

_____

7

Molecular Geometry

OBJECTIVES

- Predict molecular shapes by applying simple models of bonding and by constructing molecular models of the molecules.
- Learn about different kinds of molecular isomers.

EQUIPMENT/MATERIALS

Molecular model kit.

INTRODUCTION

Molecular Models

Speculations regarding the geometric arrangement of atoms in different molecules began with John Dalton in his atomic theory. Once the existence of atoms was accepted (which occurred much later, in the early twentieth century), visualizing their spatial arrangements was a natural next step. The spatial arrangement of the atoms of a molecule has far greater consequences than just satisfying the curiosity of chemists. The shape of a molecule and the arrangement of its constituent atoms have a direct impact on its properties. Consider enzymes and substrates. The enzyme, according to one currently accepted idea, has an active site with a specific molecular orientation into which the substrate molecule fits. Only when the substrate is held correctly in place by the enzyme does a reaction take place. A different molecule, whose atoms don't match up properly with the enzyme, won't undergo a reaction. Chemists who are trying to inhibit or enhance the activity of certain enzymes for potential treatment of particular diseases study the molecular orientation of both enzyme and substrate and use this information to design potential inhibitors or enhancers.

(a) Electron density distribution in an H–H bond. One H atom will naturally have the same electronegativity as the other. Thus the H–H bond has equal distribution of electron density and is *nonpolar*.

(b) Electron density distribution in an Li–H bond. Even though it is a smaller atom, hydrogen has a greater electronegativity than lithium and is able to attract more electron density. Thus the Li–H bond has an unequal sharing of electron density (more density around H) and is a *polar bond*.

Figure 7.1

The shape of a molecule will also determine its *polarity* (whether or not it has unequal distribution of charge). A chemical bond can be polar due to the difference in electronegativities between the two atoms that make up the bond. *Electronegativity* is defined as the ability to attract the electron density in a bond. As this difference in electronegativities increases, so does the bond polarity. (See Figure 7.1.) However, the presence of polar bonds does not necessarily mean that a molecule is polar. The geometry of the molecule may allow the individual bond polarities to "cancel out" (in a vector sense), resulting in a nonpolar molecule. Molecular polarity controls some properties of a molecule such as its solubility. (See Figure 7.2.)

The true geometry of molecules can be determined using instrumental techniques. The actual three-dimensional geometrical arrangement of the atoms in some molecules is often very complex and does not lend itself to easy visualization, either in the mind or on paper. So we resort to physical models of the molecules—something that we can physically

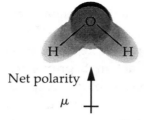

Net polarity μ

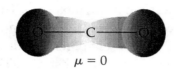

$\mu = 0$

(a) Because H_2O is a bent molecule, the polarities of the individual bonds sum up to give a net polarity as indicated.

(b) CO_2 also has polar bonds, but because the bonds are symmetrical around the central carbon atom, the individual bond polarities cancel, and the molecule is nonpolar.

Figure 7.2

construct, turn around in different orientations, sit and hold, view from different angles, and so on. There are computer programs that do this; part of your pre-lab assignment will involve manipulating molecules on a computer screen. However, it is often desirable to actually handle the model. Most molecular models use balls to represent atoms and sticks to represent the bonds between atoms. Actually, each ball represents the nucleus of an atom, but it grossly exaggerates the size of the nucleus relative to the overall size of the molecule. Similarly, the bonds in a real molecule do not look like sticks; the sticks in these models represent the line connecting the two nuclei involved in the bond. They are not an accurate representation of where the electrons in the bonds might be; the electron distribution is depicted better in the diagrams of H_2, LiH, H_2O, and CO_2 shown in Figures 7.1 and 7.2. However, space-filling models that show the electron density are often difficult to interpret when it comes to molecular geometry. When we construct ball-and-stick models, we usually try to duplicate the relative bond lengths and bond angles that will yield the "true" shape and geometry of the real molecule. Be aware of the limitations of such models and do not take them too literally, but also recognize the predictive and reasoning power that using models can provide us.

Molecular Geometries

Before we construct a molecular model, it is necessary to have a general idea of the bonding and the forces that dictate the geometrical shape of the molecule. This information can come from instrumental techniques or from theoretical models. One simple model used to predict molecular shapes is called valence shell electron pair repulsion (VSEPR). The VSEPR model proposes that the spatial arrangement of a molecule is primarily determined by the repulsive interactions among all of the valence-shell regions of electrons on each constituent atom. The valence-shell electron regions consist of both bonding and nonbonding electrons. Each electron region is regarded as being concentrated (localized) in a small area of space. (The inner filled shells do not participate in bonding and may be neglected in this approximation.) The electron regions are assumed to repel one another, so each region tends to be as far from the other regions as possible. Thus, the electron regions maximize the distance between themselves and minimize the repulsions. More detail can be found in your textbook.

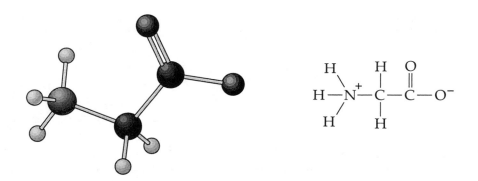

Figure 7.3   Ball-and-stick model of glycine,
which has the two-dimensional representation shown on the right.

A given number of electron regions will always assume a specific geometry—one that allows maximum distance between the regions. However, the actual shape of the molecule itself is defined by where the *atoms* are relative to one another and will be different depending on whether the electron regions are used in bonding to another atom (bonding region) or not (nonbonding electron pair). Review your class notes and your textbook and make a table of the geometries associated with the arrangement of 2–6 regions of electrons about a central atom. Then modify your table to account for situations when not all of the electron regions are bound to another atom. Your text has such a table, but reconstructing it for yourself will familiarize you with all of the possibilities and how they relate to one another other.

Isomers

Often two molecules will have the same molecular formula but different molecular orientations. When the same atoms are bonded to each other in different arrangements, they are said to be *isomers* of each other. Isomers exhibit chemical differences that reflect the different arrangements of atoms. For example, 1-propanol, 2-propanol, and methyl ethyl ether all have the formula C_3H_8O, but you can see in Figure 7.4 that they have different molecular structures. Notice also their different physical properties. There are several different kinds of isomerism. You will explore some different kinds in this exercise.

One particular type of isomerism, called *stereoisomerism*, results when the atoms have the same connectivity but different spatial orientations. There are several different kinds of stereoisomerism, but we will only explore one sort: *cis/trans* (or E/Z) isomerism. Several atoms can form multiple bonds (double or triple); among these is carbon. It forms strong double and triple bonds to itself. Although there is free rotation around a carbon–carbon single bond, a carbon–carbon double bond cannot freely rotate.[1] This gives rise to the possibility of isomers.

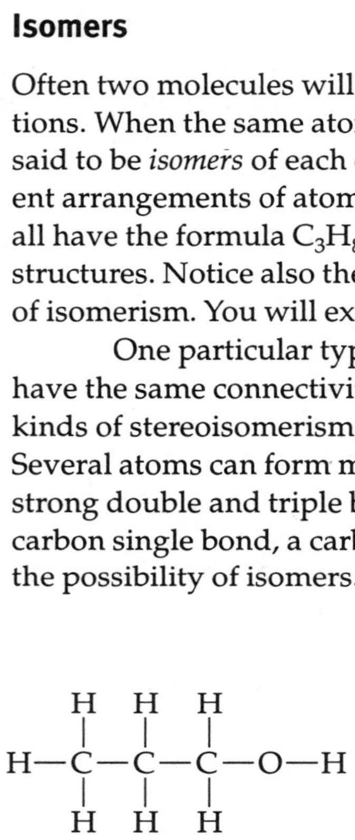

1-Propanol
bp = 97°C

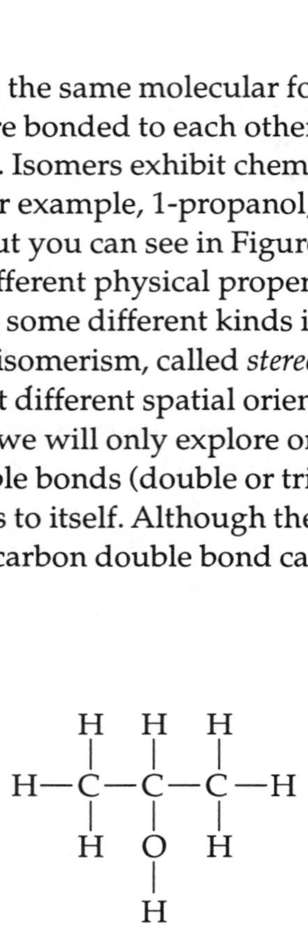

2-Propanol
bp = 82.4°C

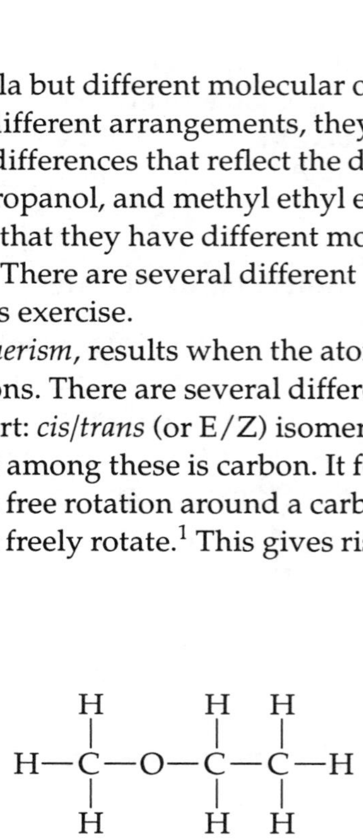

Methyl ethyl ether
bp = 10.8°C

Figure 7.4   Isomers of C_3H_8O.

[1] The reason there is no free rotation around a C=C double bond is that the bond is formed from overlap of *p*-orbitals, an aspect of bonding that is totally ignored in our simplistic VSEPR model. If the bond were to rotate, this overlap would be destroyed, thus breaking the double bond. You will learn more about this if you study more sophisticated bonding models, such as valence bond theory or molecular orbital theory.

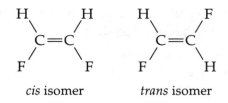

cis isomer   trans isomer

Figure 7.5   Isomers of 1,2-difluoroethene.

In the two structures of 1,2-difluoroethene in Figure 7.5 above, the atomic connectivities are the same: two carbons doubly bonded to each other, each bonded to one fluorine and one hydrogen. However, the spatial arrangement of the atoms is different in the two structures. The *cis* isomer[2] has both fluorines on the same side of the double bond, whereas the *trans* isomer has them on opposite sides. The different arrangements give rise to different physical properties: different polarities, melting and boiling points, and so on. In the exercise accompanying this lab, you will explore some *cis* and *trans* isomers. Later (in a subsequent experiment) you will explore some differences in physical properties between *cis* and *trans* compounds.

PRE-LAB EXERCISES

(The more thoroughly you do these exercises, the less time you will spend in the laboratory.)

1. Set aside an hour or two when you will have uninterrupted use of a computer with Internet access. (This will be time well spent, because it will also help you learn the material for the lecture part of the course.) Explore the following two Web sites:

   http://www.faidherbe.org/site/cours/dupuis/banque.htm#ax3

   http://www.sci.ouc.bc.ca/chem//molecule/molecule.html

   The first is a French site (don't worry, molecules are universal; you'll figure it out) designed by Gérard Dupuis. It has all the different VSEPR possibilities with numerous examples. If you have the Rasmol or CHIME plug-ins, you can rotate the molecules, but this is not necessary to derive a lot of benefit from the site.

   The second site is entitled *Molecular Models from Chemistry at Okanagan University College*, which is in British Columbia, Canada. It is designed by Dr. Dave Woodcock. He does a beautiful job of cataloging and depicting a variety of different molecules. You will need the CHIME plug-in, but the site gives you instructions on how to download it. (Having CHIME on your computer will allow you to take advantage of the vast number of quality chemistry instructional Web sites that are available.) The models on this site are of the "wire frame" variety: the bonds are depicted, but there are no round balls for the actual nuclear atoms. Essentially you will see only the bond lines. Explore as much of this site as you wish; your assignment entails scrolling down to II. Functional Groups and looking at some (not all—you won't have time!) of the molecules depicted. Pay particular attention to *alcohols*, *alkenes*, *alkynes*, and *halogenated compounds*. Then go back to the main page and

_____

[2] A good way to remember *cis* v. *trans* is that *cis* is same, and both have an "ess" sound.

scroll down to III. Molecular Type or Origin → Inorganic. Contact your instructor if you have any problems using either of these sites.

2. Begin filling out the table on the data sheet at the end of this experiment. (The central atom in each case is in boldface.) For each molecule, include the following:

- Lewis dot structure
- Total number of electron regions about the central atom
- Number of bonding electron regions
- Number of nonbonding electron pairs
- Predicted electron geometry based upon VSEPR

The Web sites listed in exercise 1 (particularly the first one) will be a great resource for helping you do this.

PROCEDURE

1. Inorganic Molecules

Make a model of each of the molecules shown in part 1 of the data sheet. On your data sheet, draw a 3-D sketch of the molecular geometry labeled with bond angles and bond polarity vectors ($\delta+$ and $\delta-$). Decide whether the molecule is polar, and if it is, indicate the direction of the molecular polarity.

2. Organic Molecules—Isomers

C_4H_{10} Build a model for butane, C_4H_{10}. Join the four carbon atoms (black balls) to each other in a "line," then add the correct number of bonds for the hydrogen atoms to the carbon atoms. You need not attach the white balls for the hydrogen atoms—there will not be enough! Just remember that there is a hydrogen atom at the end of each bond on the carbon atom. What is the geometry at each carbon atom? What is the overall shape of the molecule?

Now build 2-methylpropane, which also has the formula C_4H_{10}. Attach each of three carbon atoms to a central carbon atom. What is the geometry at each carbon atom?

This compound is known as a *branched* hydrocarbon, whereas butane is known as a *straight-chain* hydrocarbon. They are also known as *constitutional* isomers of each other. Is butane really a straight chain? Comparing these two, can you explain why one is called straight-chain and one is called branched?

C_2H_6O Build models of molecules with the formula C_2H_6O. How many isomers are there? Sketch them. Would you expect the properties of the isomers to be the same or different?

$C_2H_2Cl_2$ Draw as many Lewis dot structures for dichloroethene, $C_2H_2Cl_2$, as you can. Now build models of them. How many isomers are there? For one of the Lewis dot structures, there are actually two isomers. What are they? Which is *cis* and which is *trans*? Why are there no *cis* and *trans* versions for the other isomer?

C₂HClBrF   Draw as many Lewis dot structures for $C_2HClBrF$ as you can. (*Hint*: Start with a C=C and attach the other atoms.) Make models of them. How many isomer possibilities are there? You will encounter a problem assigning *cis* and *trans* designations to these isomers. What is that problem? Can you invent a way of naming these isomers that gets around this shortcoming of the *cis* and *trans* way of naming double-bond isomers? Describe this method.

Data Sheet for Unit 7

Name _____
Date _____

1. Inorganic Molecules

Formula	Lewis dot structure	Number of electron regions	Number of bonding regions	Number of nonbonding regions	Predicted geometry	Sketch of molecular structure	Polarity
CF_2Cl_2							
ClF							
OCl_2							
BF_3							
H–C–C–H							

Data Sheet for Unit 7

Name_____

Date_____

1. Inorganic Molecules (*continued*)

Formula	Lewis dot structure	Number of electron regions	Number of bonding regions	Number of nonbonding regions	Predicted geometry	Sketch of molecular structure	Polarity
NH_3							
CO_2							
NH_4^+							
$SeCl_2$							
H_2CO							

1. Inorganic Molecules (*continued*)

Formula	Lewis dot structure	Number of electron regions	Number of bonding regions	Number of nonbonding regions	Predicted geometry	Sketch of molecular structure	Polarity
NO_2^-							
BeH_2							
SiO_2							
SO_3							
CO_3^{2-}							

Data Sheet for Unit 7

Name _____

Date _____

1. Inorganic Molecules (*continued*)

Formula	Lewis dot structure	Number of electron regions	Number of bonding regions	Number of nonbonding regions	Predicted geometry	Sketch of molecular structure	Polarity
H_2O							
H_3O^+							

2. Organic Molecules—Isomers

C_4H_{10}

1. Butane. What is the geometry at each carbon atom? What is the overall shape of the molecule? Draw the structure below. Is it really a straight chain? Explain.

_____

_____

_____

2. 2-Methylpropane. What is the geometry at each carbon atom? Draw the structure below.

3. Comparing these two, can you explain why one is called straight-chain and one is called branched?

_____

_____

_____

_____

C_2H_6O

How many isomers are there? Sketch them. Would you expect the properties of the isomers to be the same or different?

_____

_____

_____

_____

$C_2H_2Cl_2$

1. Draw the Lewis dot structures below.

2. How many isomers are there? Sketch them below. Which Lewis dot structure has two isomers? Which is *cis* and which is *trans*? Why are there no *cis* and *trans* versions for the other isomer?

_____

_____

_____

_____

$C_2HClBrF$

1. Draw the Lewis dot structures below.

2. How many isomer possibilities are there? Draw them below.

3. Explain the difficulty in assigning *cis* and *trans* designations to the isomers you drew above. Can you invent a way of naming these isomers that gets around this shortcoming of the *cis* and *trans* way of naming double-bond isomers? Describe this method.

_____

_____

_____

_____

_____

_____

_____

_____

_____

_____

_____

_____

_____

_____

_____

_____

_____

_____

8

Analysis and Identification of Two Geometric Isomers

OBJECTIVES

- Compare melting point and solubility results with literature values to identify two isomers.
- Isomerize a *cis* isomer into its *trans* counterpart through chemical reaction.

EQUIPMENT/MATERIALS

Samples of ethyl alcohol, dimethyl ether, maleic acid, fumaric acid, unknowns A and B, test tubes, stirring rod, flasks, distilled water, concentrated HCl, funnel, filter paper, watch glass.

INTRODUCTION

When two chemicals have the same molecular formula but different molecular structures, we call them *isomers*. Ethyl alcohol and dimethyl ether are isomers. They both have the formula C_2H_6O, but as the Lewis dot structures in Figure 8.1 illustrate, the atoms are arranged differently. As you can see, this difference in arrangement leads to a significant difference in

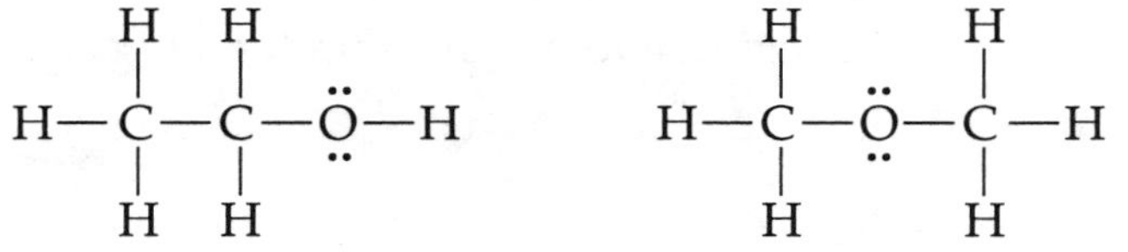

Ethanol
mp = −114°C
bp = 78°C

Dimethyl ether
mp = −141°C
bp = −24.8°C

Figure 8.1   Lewis dot structures of isomers with the formula C_2H_6O.

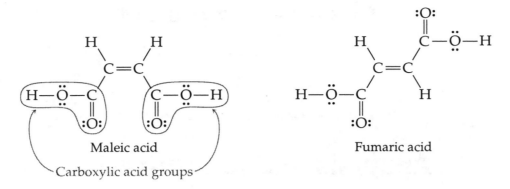

Figure 8.2   Lewis dot structures of maleic and fumaric acid.

properties. For example, dimethyl ether is a gas at room temperature, whereas ethanol doesn't boil until the temperature reaches 78°C. These differences in properties can be attributed to differences in molecular geometry.

Let's look at two other isomers, both with the formula $C_4H_4O_4$. Their Lewis dot structures are shown in Figure 8.2. Both maleic acid and fumaric acid have two carboxylic acid groups. However, in maleic acid these two groups are on the same side of the double bond, whereas in fumaric acid they are on opposite sides of the double bond. Carbon–carbon double bonds do not freely rotate. Thus these two isomers are chemically different and have different properties. Does either of these molecules have a dipole moment? In maleic acid, the two carboxylic acid groups are said to be *cis*, whereas in fumaric acid, the groups are said to be *trans*. (An easy way to remember is that *cis* and *same* start with the same "*ess*" sound.)

The two *cis* carboxylic acid groups are rather crowded in maleic acid, and for this reason it has the higher energy of the two structures. It is often possible to convert a crowded *cis* isomer to the more stable *trans* form by a chemical reaction. This process, called *isomerization*, requires an input of energy to rupture the double bond and allow free rotation of the less crowded isomer. After rotation, the double bond is reformed. A similar reaction occurs in your body when certain chemicals in your eye, which are responsible for vision, respond to light by undergoing an isomerization reaction.

In this experiment, you will receive unspecified samples of maleic acid and fumaric acid. Your job is to use their properties and what you know about the relationship between molecular structure and properties to figure out which is which. Once you have that sorted out, you will attempt to isomerize maleic acid into fumaric acid, using heat as your source of energy along with hydrochloric acid, which acts as a catalyst. (A catalyst is a substance that speeds up a chemical reaction without itself undergoing a permanent chemical change in the process.)

PRE-LAB QUESTIONS

1. Look up maleic and fumaric acids, either on the Web or in a chemistry reference book (such as the *CRC*, the *Merck Index*, or the *Aldrich Catalog*). Your instructor can give you guidance in doing this. Record the relative melting points of the two substances in your

notebook. Which has the higher melting point? Postulate as to how the symmetry of the molecule might affect its melting point.

2. Does either maleic or fumaric acid have a dipole moment? Draw structures of the molecules and indicate any dipole moments that are present. Which would you expect to be more soluble in a polar solvent like water? Explain your reasoning.

PROCEDURE

1. Analysis by Melting Point Comparison

On the reagent bench you will find two jars. One is labeled A and the other B. Label two small test tubes A and B as well. Place a pea-sized portion of each sample (A and B) in its corresponding test tube.

Light a Bunsen burner. Hold each test tube with a separate test-tube holder in the same hand. Heat the samples simultaneously until one of them melts. Keep the bottoms of both test tubes in the same position in the flame. Record on your data sheet which sample has melted. Allow the test tubes to cool before cleaning them.

2. Analysis by Comparing Solubilities

Weigh out 0.5 g each of A and B. Place each sample in a test tube. Add 10 mL of distilled water and stir gently with a stirring rod. Record the relative solubilities of A and B in your notebook.

3. Isomerization of Maleic Acid

From the results of the melting point and solubility analysis you performed, decide which of the two samples, A or B, is maleic acid.

Weigh out 1.0 g of maleic acid. Place it in a 100-mL Erlenmeyer flask. Add 10 mL of distilled water to the flask and swirl gently to dissolve.

In the hood, heat the flask on a hot plate until the water is boiling. Measure approximately 10 mL of concentrated hydrochloric acid in a 100-mL beaker. Then measure out *exactly* 10 mL of concentrated HCl in a 10-mL graduated cylinder. (The two-step process is used to prevent spills; it is easier to pour from the big bottles into a beaker than into the narrow mouth of a graduated cylinder.)

CAUTION: CONCENTRATED ACIDS CAN CAUSE SEVERE CHEMICAL BURNS. IF YOU SPILL AN ACID ON YOURSELF, WASH THE CONTAMINATED AREA THOROUGHLY WITH SATURATED BICARBONATE SOLUTION (AVAILABLE ON TOP OF THE LAB BENCHES).

Add the concentrated HCl *slowly* and *cautiously* to the boiling water in the flask and continue to boil for another minute. Record your observations. Turn off the hot plate and use tongs to remove the flask from the hot plate.

Allow the flask to cool on the bench for at least 10 minutes. (Do not cool in an ice bath yet because the reaction may still be proceeding.) Continue to record your observations. Cool 10 mL of distilled water in a small clean test tube in an ice bath.

Cool the reaction flask in the ice bath. Filter and collect the solid that is formed using gravity filtration. Spread the solid out on a watch glass to dry.

Test the solubility of a pea-sized portion of your product and compare it to the results in the solubility analysis (part 2) you performed above. You may need to repeat the solubility test from part 2 in order to get a direct comparison. Record your observations on your data sheet.

Data Sheet for Unit 8

Name_____

Date_____

1. **Analysis by Melting Point Comparison**

   Which sample (A or B) melted first? _____

2. **Analysis by Comparing Solubilities**

   Which sample (A or B) is more soluble in water? _____

3. **Isomerization of Maleic Acid**

   Which sample is maleic acid? _____

   Which sample is fumaric acid? _____

   Briefly explain your reasoning.

   _____

   _____

   _____

   _____

   _____

   _____

   _____

   Describe the results of the solubility test of the sample of maleic acid you heated with acid. Did a reaction occur? If so, what was it?

   _____

   _____

   _____

   _____

   _____

   _____

   _____

   _____

   _____

   _____

   _____

   _____

   _____

   _____

9

Stoichiometry— Gravimetric Analysis of a Gas-Forming Reaction

OBJECTIVES

- Explore the relationship between a macroscopic phenomenon (mass loss) and a molecular relationship (balanced chemical equation).
- Learn how to determine number of moles from concentration information.
- Learn how to organize and manage numerous experimental measurements.

EQUIPMENT/MATERIALS

Sodium bicarbonate, sodium carbonate, dilute acid, beakers, graduated cylinder, balance.

INTRODUCTION

The title of this experiment sounds very impressive: Gravimetric analysis? Gas-forming reactions? Stoichiometry? What does it all mean? The actual experiment is very simple, as is the method of analysis. It is simple enough that you could potentially do it at home in your kitchen. Simply stated, you will be monitoring the progress of a reaction by keeping track of changes in mass.

How can mass change? We know that matter cannot be created or destroyed by a normal chemical process. This is where the second part of the title is important: because the reaction forms a gas and the system is open to the atmosphere, the gas can diffuse away and thus the system will lose mass as the reaction progresses. Can you think of an example in which a reaction system might *gain* mass as a reaction progresses?

It is very likely that you have already observed at least one of the reactions that you will be studying in this experiment. Reacting baking soda and vinegar produces much foaming and is a popular way for children to explore chemistry by making messes in the kitchen. It involves a simple reaction between sodium bicarbonate (baking soda) and an acid (in this case vinegar):

$$NaHCO_3(s) + H^+ \longrightarrow Na^+ + CO_2(g) + H_2O(l)$$

Sodium   Acid
bicarbonate

You will use HCl as your acid, so the overall reaction can be written as follows:

$$NaHCO_3(s) + HCl(aq) \longrightarrow NaCl(s) + CO_2(g) + H_2O(l)$$

The other reaction used in this experiment is very similar but uses sodium carbonate (Na_2CO_3) rather than sodium bicarbonate:

$$Na_2CO_3(s) + 2\,HCl(aq) \longrightarrow 2\,NaCl(s) + CO_2(g) + H_2O(l)$$

PRE-LAB EXERCISES

1. Calculate the molar mass of $NaHCO_3(s)$, $Na_2CO_3(s)$, and $CO_2(g)$.

2. Calculate the number of moles of HCl in the two volumes of HCl used in the procedure.

3. In this experiment, you will prepare your own data tables with some guidance provided below.

   First, prepare a table that lists the reagents, their molar masses, the amounts you'll use in the experiment and the number of moles that represents, and any properties or safety issues that are important for you to know about as you work in the lab. This table should be completed before you start your experiment.

   Second, prepare two tables that you will use to record data during the experiment—one for the $NaHCO_3$ runs (procedure 1) and one for the Na_2CO_3 runs (procedure 2). Each table should have spaces to record data for two runs. These tables should contain the following categories *for each run*:

   a. Mass of beaker (measure)

   b. Mass of beaker + $NaHCO_3$ (measure)

   c. Mass of $NaHCO_3$ (calculate: line b − a)

   d. Mass of HCl solution (calculate using volume and density)

   e. Total initial mass of reagents, HCl solution + $NaHCO_3$ (calculate: c + d)

   f. Final mass of reagents + beaker at end of reaction (measure)

   g. Final mass of reagents (f − a)

   h. Total mass lost during reaction (e − g)

   The third table will have Na_2CO_3 rather than $NaHCO_3$, but the rest will be analogous. Leave space in your notebook for writing observations for the two procedures.

PROCEDURE

1. NaHCO₃(s) + HCl(aq)

Obtain two 400-mL beakers and label them "Run 1" and "Run 2." Determine their masses and record these values in your notebook in the appropriate data table. Obtain a 100-mL graduated cylinder.

Into beaker 1, weigh out approximately 1.3 g of $NaHCO_3(s)$, but *determine and record the exact mass* in your data table.

Place 25.0 mL of 1.0 M HCl into the graduated cylinder. Assume the density of this solution is 1.0 g/mL and calculate the mass of the HCl solution. Write this number in your notebook and add it to the mass of the $NaHCO_3(s)$ to determine the total mass of the reagents. Record this value in your notebook.

Carefully add the HCl solution to the contents of beaker 1 (containing $NaHCO_3$).

> **CAUTION: THE REACTION WILL CAUSE CONSIDERABLE FOAMING. YOU WILL NEED TO POUR *SLOWLY* AND/OR INTERMITTENTLY SO THE FOAMING CAN SUBSIDE. IF THE REACTION MIXTURE SPLASHES OVER THE WALLS OF THE BEAKER, YOU WILL HAVE TO CLEAN IT UP AND START OVER.**

When the foaming has subsided, swirl the beaker gently to free any bubbles that are adhering to the walls of the beaker. Measure the mass of the beaker after the reaction is completed.

Repeat this process (Run 2). Carefully rinse and dry your beakers between runs.

2. Na₂CO₃(s) + HCl(aq)

Repeat this process using Na_2CO_3 and HCl(aq), making the changes indicated below.

Into beaker 1, weigh out approximately 1.3 g of $Na_2CO_3(s)$, but *determine and record the exact mass* in your notebook.

Place 35.0 mL (note different volume) of 1.0 M HCl in the graduated cylinder. Assume the density of this solution is 1.0 g/mL and calculate the mass of the HCl solution. Write this number in your data table and add it to the mass of the $Na_2CO_3(s)$ to determine the total mass of the reagents. Record this value in your data table.

Carefully add the HCl solution to the contents of beaker 1 (containing Na_2CO_3).

> **CAUTION: AS DESCRIBED ABOVE, FOAMING IS A PROBLEM. YOU WILL NEED TO POUR SLOWLY AND/OR INTERMITTENTLY SO THE FOAMING CAN SUBSIDE. IF THE REACTION MIXTURE SPLASHES OVER THE WALLS OF THE BEAKER, YOU WILL HAVE TO CLEAN IT UP AND START OVER.**

When the foaming has subsided, swirl the beaker gently to free any bubbles that are adhering to the walls of the beaker. Measure the mass of the beaker after the reaction is completed.

Repeat this process (Run 2). Carefully rinse and dry your beakers between runs.

REPORT

1. Prepare another table with the following headings (either across or down):

   - Run[1]
   - Substance[2]
   - Mass[3]
   - Moles[4]
   - Theoretical moles of CO_2 lost[5]
   - Theoretical mass of CO_2 lost[6]
   - Actual mass of CO_2 lost[7]
   - Percent error[8]

   Perform the calculations necessary to determine the values in the table. Guidance is given below.

   Provide a single sample calculation for each category that illustrates where you got the numbers you put in the table.

2. Demonstrate mathematically that the HCl is in excess in these reactions.

3. How well do the experimental data match the theoretical expectations? Postulate as to some causes for potential discrepancies.

---

[1] Run 1 or Run 2

[2] Either $NaHCO_3$ or Na_2CO_3

[3] Measured in your experiment

[4] Moles = Mass/Molar mass

[5] For $NaHCO_3(s) + HCl(aq) \longrightarrow NaCl(s) + CO_2(g) + H_2O(l)$, moles of $NaHCO_3$ = moles of $CO_2(g)$.

For $Na_2CO_3(s) + 2\,HCl(aq) \longrightarrow 2\,NaCl(s) + CO_2(g) + H_2O(l)$, moles of Na_2CO_3 = moles of $CO_2(g)$.

[6] Mass of CO_2 = moles $\times$ 44.01 g/mol

[7] Measured in your experiment

[8] Percent error $= \dfrac{|\text{Theoretical value} - \text{Experimental value}|}{\text{Theoretical value}} \times 100$

10

Heats of Combustion and the Caloric Content of Food

OBJECTIVES

- Use calorimetry and combustion to determine the caloric content of different types of nuts.
- Correlate experimental results with nutritional information on commercial labels and account for discrepancies.

EQUIPMENT/MATERIALS

Assorted nuts, 2 250-mL beakers, tongs, Bunsen burner, modeling clay or Play-Doh®, water, small nail or pin, thermometer, 100-mL graduated cylinder, wire gauze, ring clamp, ring stand.

INTRODUCTION

Nutritional labels on the foods we eat provide a great deal of information. One piece of information many of us notice in particular is the caloric content. We all know that cookies and potato chips contain greater amounts of calories than healthier alternatives such as fruits and vegetables. You probably also know that if you eat more calories than you expend in metabolic processes and exercise, you will gain weight. If you consume fewer calories than you expend, you will lose weight. However, if you look at the scientific definition of a calorie, you will see it defined as the amount of heat required to raise the temperature of 1.000 g of water from 25°C to 26°C. What does this have to do with food?

To understand the relationship between calories and food, we must look at heat and energy from the perspective of a chemical reaction. When chemical bonds break and re-form in a chemical reaction, energy is often absorbed or released. A good example is photosynthesis. Green plants use energy from the Sun and process this energy through a molecule called chlorophyll to convert water and carbon dioxide into glucose (sugar) and oxygen. In reality there are numerous, complicated steps, but the overall reaction is summarized in Figure 10.1.

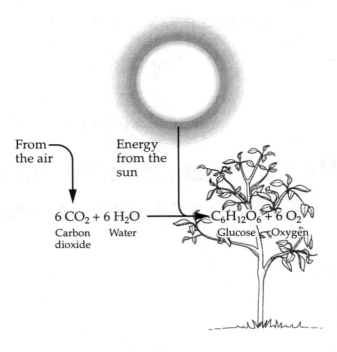

From the air

Energy from the sun

$6\ CO_2 + 6\ H_2O$ → $C_6H_{12}O_6 + 6\ O_2$

Carbon dioxide   Water   Glucose   Oxygen

Figure 10.1   Photosynthesis.

The plant then stores and metabolizes this glucose and recoups that energy, using it for various metabolic processes, such as growth, transpiration, and reproduction. Furthermore, any organism that eats the plant will also be able to get energy from the glucose present. The reaction for the metabolism of glucose is ultimately just the reverse of photosynthesis:

$$C_6H_{12}O_6 + 6\ O_2 \longrightarrow 6\ CO_2 + 6\ H_2O + \text{Energy for metabolism (heat)}$$

Interestingly, the overall energy associated with a chemical reaction is the same whether the reaction occurs stepwise at low temperature inside a living cell or rapidly at high temperature inside a furnace. In this case, the oxidation of glucose that occurs inside a cell and is regulated by various enzymes is nothing more than the combustion of glucose. So the energy released gradually inside the cell is overall the same energy that would be released if you simply burned glucose. In the cell it is just released in small manageable increments. (See Figure 10.2.)

So what does all this have to do with the calories? Well, let's think about how we measure and use heat. Think about what occurs when you turn on the burner underneath a pot of water on your stove. If you have a gas range, you are burning methane (CH_4) to produce carbon dioxide and water. Just like the metabolism of glucose, this reaction gives off heat.

$$CH_4 + 2\ O_2 \longrightarrow CO_2 + 2\ H_2O + \text{Heat}$$

The heat released by this reaction is absorbed by the pot and the water it contains. As the pot–water combination absorbs this heat, the temperature rises. (See Figure 10.3.) The greater the amount of water, the more heat is required to raise it to a certain temperature. A simple way to measure the heat used and released by different chemical reactions is to connect it to the change in temperature of a particular substance, like water. This is how a calorie is defined. As mentioned above, it is the amount of heat that is required to raise the temperature of water 1 C°.

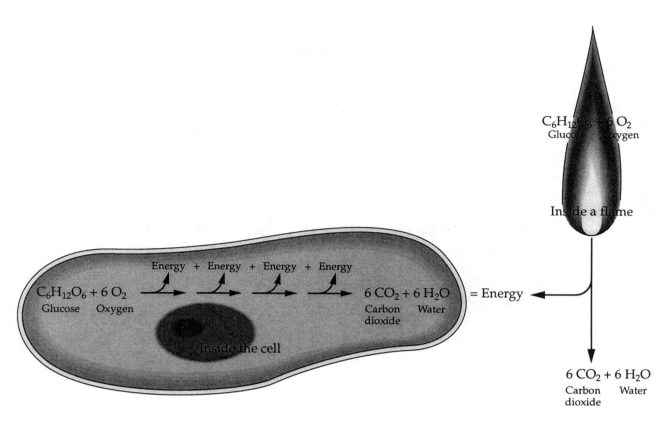

Figure 10.2   Metabolism of glucose.

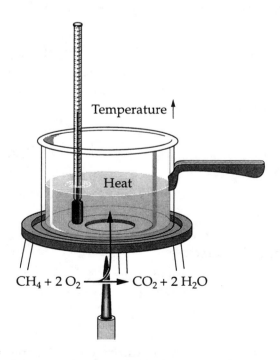

Figure 10.3   Heating water on a gas burner.

We talk about consuming calories, but how can we eat heat? As we digest and metabolize the food that we eat, bonds are broken and re-formed. These processes release heat that your body can use to fuel other processes that require heat. The energy derived from food is described by its caloric content. However, it is important to note the difference between the calorie and the Calorie with an uppercase C. The calorie is defined above (the amount of heat required to raise the temperature of water 1 C°). The Calorie is equal to 1000 "little c" calories. Another way to state this is that a Calorie is actually a kilocalorie or kcal. So when you eat a bagel that contains 280 Calories (280 kcal), in reality it provides 280,000 calories of energy, enough to raise the temperature of 280 kg of water 1 C°.

In most chemistry textbooks, the heat associated with chemical processes is expressed in joules (J) or kilojoules (kJ). There are 4.184 J in 1.00 calorie. Thus, that 280-Calorie bagel contains (4.184 J/cal)(280,000 cal) = 1,172,000 J: over 1 million joules! And that's without the cream cheese!

In this experiment, you will be relating the caloric content of foods, specifically nuts, to the energy required to heat water. The experiment requires a flame, so you should exercise caution. You will use the heat capacity of water in calories and its temperature change to determine the caloric content. The equation you will use to accomplish this is

$$q \text{ (cal)} = C_{\text{water}} \times \Delta T$$

where

q is the heat associated with the process

C_{water} is the heat capacity of water in calories (with a little c) rather than joules

$C_{\text{water}} = 1.0 \text{ cal/g} \times \text{Mass of water in grams}$

ΔT is the change in temperature $= T_{\text{final}} - T_{\text{intial}}$ in C°

We will assume the density of water is ~1 g/mL, thus 100 mL of water ≈100 g of water.

You will also need to determine the caloric content given on the label for comparison to your experimentally derived number. Remember the calories given are actually kilocalories, and as serving sizes may be given in ounces, you may need the following conversion factor: 28.35 g = 1 ounce.

PRE-LAB EXERCISE

1. How many calories are used to heat a 50.00-g sample of water from 8.1°C to 27.4°C? How many kilocalories?

2. At the top of the next page is the label from a jar of almonds. Determine the number of kilocalories per gram of almonds.

3. Prepare data tables for the experiment. Your instructor will tell you how many nuts you are expected to analyze and how many trials per nut you will be expected to perform. Your data tables should include the following headings:

Type of nut	Mass of nut	Initial water temperature (°C)	Final water temperature (°C)	Comments

Nutrition Facts

Nutrition Facts
Serving Size 1 ounce (30 g)
Servings Per Container 1

Amount Per Serving
Calories 170 Calories from Fat 140

	% Daily Value *
Total Fat 15 g	23 %
Saturated Fat 1 g	5 %
Polyunsaturated Fat 4 g	
Monounsaturated Fat 10 g	0 %
Cholesterol 0 mg	0 %
Sodium	
Total Carbohydrate 5 g	2 %
Dietary Fiber 4 g	8 %
Sugars 1 g	
Protein 6 g	
Vitamin A 0 %	Vitamin C 0 %
Calcium 8 %	Iron 6 %
Vitamin E 35 %	Folate 4 %
Magnesium 22 %	Phosphorous 14 %

*Percent daily values are based on a 2,000 calorie diet.

PROCEDURE

Use a graduated cylinder to measure 100 mL of cold tap water. Pour the water into a 250-mL beaker.

Obtain a nut sample and determine and record its mass. If the mass is greater than 2–3 g, you will need to choose a smaller nut or break off a smaller piece to use in the experiment. A nut or piece of nut 1 g or less will work well.

Place the nut firmly onto the end of the pin or nail. If the nut splinters, you will need to start over with a new piece of nut. Wrap the other end of the pin with a cone made of modeling clay and secure it to the bottom of the inverted beaker. (See Figure 10.4.)

Attach the ring clamp to a ring stand. Place the wire gauze on the ring clamp and adjust the height of the clamp and the position of the inverted beaker so that the nut is as close to the wire gauze as possible (without touching).

Measure and record the initial temperature of the water.

Ignite the nut with a Bunsen burner. It may take repeated efforts to get it lit, so you will want to avoid using smaller ordinary matches. Place the thermometer in the water (ensuring that it is not touching the bottom of the beaker) and note the temperature rise inside the water.

Continue to heat the water with the burning nut, paying attention to the rising temperature.

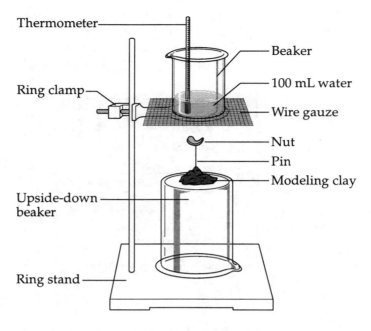

Figure 10.4   Experimental setup.

When the temperature stops rising and/or the nut extinguishes, record the final highest temperature.

Perform as many trials per type of nut and examine as many different kinds of nuts as instructed. Start each trial with a fresh sample of cold water. Your instructor may also have other nutlike foods (e.g. "wheat nuts") available for your analysis.

Record the following information in your notebook or on your data sheet from the nutritional labels of the various nuts or nut substitutes you tested:

- Type of nut
- Mass per serving
- Calories (kcal) per serving

REPORT

1. Prepare a new table (or a series of tables) that provides the following information. You may wish to simply extend the data table you used to record the results of the experiment.

    - Type of nut
    - Mass of nut sample
    - Change in temperature ($\Delta T = T_f - T_i$)

- Experimental caloric content per gram, $q_{per\ gram}$:

$$q_{sample}\ (cal) = 1.0\ cal/g \times \text{Mass of water in grams} \times \Delta T$$

$$q_{per\ gram} = \frac{q_{sample}}{\text{Mass of nut sample (g)}}$$

- Label-derived caloric content per gram; see example calculation below.

2. Using a graphing program, graph your results in bar-graph form. Place the experimental average kcal/g next to the label value for each type of nut.

3. How do the caloric content values that you determined experimentally compare with those reported on the labels of the different kinds of nuts?

4. The heat transfer from the combustion of the nut to the increase in water temperature was not perfect. List all the places where heat could be lost in this process.

5. How could you design an experimental apparatus that would minimize the loss of heat to the surroundings?

Example Calculation for Determining the Label-Derived Caloric Content per Gram

Use the label to determine the number of kilocalories per ounce and convert that to kcal/g. If the label says that there are 325 Calories per serving and a serving is 2 oz, use the ounce-to-gram conversion, 28.35 g = 1 ounce, to determine that

$$2\ oz \times \frac{28.35\ g}{1\ oz} = 56.7\ g$$

Then set up the ratio

$$\frac{325\ kcal}{56.7\ g} = \frac{?\ kcal}{1.0\ g}$$

to determine that there are about 5.7 kcal/g.

Don't forget about the Calorie/calorie discrepancy. Labels provide information about Calories, which are in reality kilocalories, where 1000 cal = 1 kcal. The measurements and calculations you will make in this experiment will provide you with numbers of calories (with a little c).

11

Redox Reactions

OBJECTIVES

- Explore a redox activity series.
- Prepare an electrochemical cell.

EQUIPMENT/MATERIALS

Zinc, copper, and magnesium strips, solutions of potassium nitrate, zinc sulfate, copper sulfate, and magnesium sulfate, paper towel or filter paper, sandpaper, beakers, test tubes, alligator clips, wire, electronic device.

INTRODUCTION

Oxidation–reduction or *redox* reactions turn up in just about every aspect of your life. When you drive a car or ride a bus, the burning of gasoline or other fuel is the redox reaction that gets you where you are going. When you ride a bike or walk, your body uses a very complex redox system to convert your fuel (food) into energy. The rusting of metal, bleaching of stains on clothes or teeth, electroplating of metals, and the operation of a battery are all examples of redox reactions.

A redox reaction involves a transfer of electrons from one species to another. Let's look at a simple example, the dissolution of zinc metal by hydrochloric acid. Remember that aqueous hydrochloric acid, $HCl(aq)$, is dissociated into $H^+(aq)$ and $Cl^-(aq)$ ions; it is the hydrogen ions that are reacting in this example. The overall reaction is shown below.

$$2\,H^+(aq) + Zn(s) \longrightarrow H_2(g) + Zn^{2+}(aq)$$

Let's look at what's happening to each of the reactants. The $Zn(s)$ is being converted to $Zn^{2+}(aq)$. It is losing 2 electrons and being converted from a neutral atom to a positively charged ion. We say that it is being *oxidized*; *oxidation* is the *loss* of electrons. Looking only at the zinc, we could write this reaction:

$$Zn(s) \longrightarrow Zn^{2+}(aq) + 2\,e^-$$

where the 2 e$^-$ represents the 2 electrons that the zinc atom lost. This is called a *half reaction* because it is just half of what is going on.

The two H$^+$ ions are being converted to a neutral H$_2$ molecule. They have gained the 2 electrons that the zinc metal atom has lost. The H$^+$ is thus *reduced* in this reaction. (Its charge has certainly been reduced from +1 to 0.) *Reduction* is the *gain* of electrons. That half reaction can be written as:

$$2 \, H^+(aq) + 2 \, e^- \longrightarrow H_2(g)$$

If we put those two half reactions together, the electrons will cancel and we'll wind up with the reaction we have written above:

$$Zn(s) \longrightarrow Zn^{2+}(aq) + 2\,e^-$$
$$\underline{2 \, H^+(aq) + 2\,e^- \longrightarrow H_2(g)}$$
$$2 \, H^+(aq) + Zn(s) \longrightarrow H_2(g) + Zn^{2+}(aq)$$

Just like milk and cookies or love and marriage, you cannot have oxidation without reduction and vice versa. In order for something to be oxidized, something else must be reduced. The electrons have to be transferred from one species to another, and the number of electrons has to balance. You can't have the atom or molecule that is being reduced gaining more (or fewer!) electrons than the oxidized atom or molecule is giving up.

In this experiment, you will explore two aspects of redox reactions. The first is an activity series and the second involves making an electrochemical cell (battery).

Activity Series

The *activity* of a metal is defined as its susceptibility to oxidation. The more active a metal is, the more likely it is to be oxidized. Less active metals are less prone to oxidation. Thus, a more active metal will react with the metal ion of a less active metal. For example, sodium is more active than copper, so the following reaction will occur:

$$Na(s) + Cu^+(aq) \longrightarrow Na^+(aq) + Cu(s)$$

It follows that if sodium is more active than copper, then copper is less active than sodium, so the following reaction does *not* occur:

$$Na^+(aq) + Cu(s) \xrightarrow{\hspace{0.3cm}\times\hspace{0.3cm}} Na(s) + Cu^+(aq)$$

In this experiment, you will explore some metals and determine their order of redox activity.

Electrochemical Cells

Electrochemical cells use metals (and other substances) of different activity to produce an electric current. The oxidation and reduction reactions take place apart from each other, so the transfer of electrons happens through a wire. The energy associated with this process can be used to do work—for example, play a radio, start a car, or do any of the other things we use batteries for. Getting the reactions to happen apart from each other is tricky. A typical electrochemical cell is shown in Figure 11.1.

The reduction reaction is occurring on the right, where silver ion is being reduced to silver metal. The oxidation of copper is occurring on the left, where copper metal is being

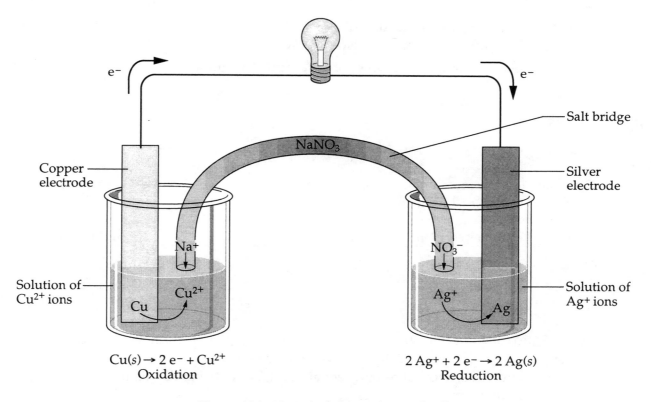

$$Cu(s) \rightarrow 2\,e^- + Cu^{2+}$$
Oxidation

$$2\,Ag^+ + 2\,e^- \rightarrow 2\,Ag(s)$$
Reduction

Figure 11.1   A typical electrochemical cell.

oxidized to copper ion. The electrons are being transferred from the copper to the silver through the wire that connects the two half reactions. The other piece of the electrochemical cell is the *salt bridge*. The salt bridge allows ions to flow so that the charges stay balanced. Without the salt bridge, the whole process would screech to a halt because you cannot have negative ions running around a solution without positive ions to balance them. (Nor can you have the opposite: positive ions without their charges balanced by negative ions.) The electric current generated from this reaction is being used to do work (the light bulb). In this experiment, you will set up an electrochemical cell and use it to do work.

PRE-LAB QUESTIONS

The following results were obtained through a series of experiments:

Reactions with acids:

	Cr(s)	*Ag(s)*
HCl	*Metal dissolves, bubbles, green solution*	*No apparent reaction*
H_2SO_4	*Metal dissolves, bubbles, bluish purple solution*	*No apparent reaction*

Replacement reactions:

Cr^{2+} + Ag metal: no apparent reaction, silver metal did not change, solution stayed green.

Ag^+ + Cr metal: silvery solid formed on piece of Cr metal. Solution went from colorless to bluish green.

1. Write net ionic equations that describe each of the results noted above. Explain the color change observed in the second replacement reaction.

2. Briefly discuss the relative activities of silver and chromium.

PROCEDURE

1. Activity Series

Copy the matrix below into your notebook. The numbers correspond to the numbering system for the test tubes. (See instructions below.)

	Initial observations	Zn metal	Cu metal	Mg metal
$ZnSO_4$		1 _____	2 _____	3 _____
$CuSO_4$		4 _____	5 _____	6 _____
$MgSO_4$		7 _____	8 _____	9 _____

Obtain a strip of zinc foil, a strip of copper foil, and a strip of magnesium. Carefully sand each strip lightly with the sandpaper provided to expose a fresh metal surface. Break each metal strip into three pieces of roughly equal size. (Try not to handle them too much with your greasy, grubby fingers!)

Obtain nine small test tubes and label them 1–9.

Place ~1–2 mL of the 1.0 M zinc sulfate solution in test tubes 1–3.

Place ~1–2 mL of the 1.0 M copper sulfate solution in test tubes 4–6.

Place ~1–2 mL of the 1.0 M magnesium sulfate solution in test tubes 7–9.

Record the initial colors and other identifying characteristics of the solutions in your matrix.

Place a zinc strip in test tubes 1, 4, and 7. Place a copper strip in test tubes 2, 5, and 8. Place a magnesium strip in test tubes 3, 6, and 9.

Make observations of each of the nine test tubes. Record your observations in your matrix.

2. Electrochemical Cell

Obtain two 100-mL beakers. Place ~50 mL of a 1.0 M solution of zinc sulfate ($ZnSO_4$) in one and ~50 mL of a 1.0 M solution of copper sulfate ($CuSO_4$) in the other. Place a 1×10 cm strip of zinc foil in the zinc sulfate solution and a 1×10 cm strip of copper foil in the copper sulfate solution.

Obtain a paper towel (or filter paper) and roll it into a tight cylinder. Place it in a beaker of 1.0 M potassium nitrate (KNO_3) solution until it is thoroughly soaked. Place the two beakers containing the zinc and copper strips very close to each other. Remove the paper towel, bend it in the middle, and insert it into both beakers so that it connects them. (See Figure 11.2.) This will act as your salt bridge. Make sure the paper is at least 1 cm beneath the surface of each solution.

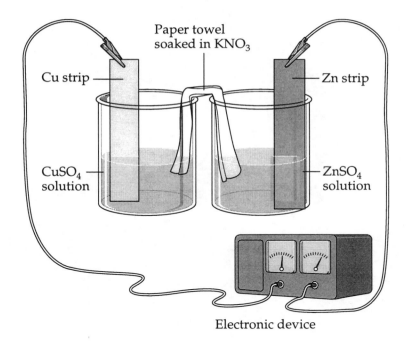

Paper towel
soaked in KNO$_3$

Cu strip —

Zn strip

CuSO$_4$
solution

ZnSO$_4$
solution

Electronic device

Figure 11.2   Experimental setup.

Connect the zinc and copper foils to wires using alligator clips. Connect the alligator clips to the device indicated by your instructor (typically a voltmeter, light bulb, small electronic device, or LED). Record your observations of the two solutions, the foil strips, and the device as the reaction progresses.

Carefully remove the salt bridge. What happens?

Replace the salt bridge. Continue to make observations.

REPORT

1. **Activity Series**

   a. Turn in a copy of your matrix that includes all your observations.

   b. Write reactions for what did or did not occur in each test tube. (For example, if a reaction did not occur, write the reagents present and NR after the arrow.)

   c. Rank the metals in order of increasing activity. In a brief paragraph, support your ranking.

2. **Electrochemical Cell**

   a. Write the reaction(s) that occurred in your cell.

   b. Was there enough current generated by the battery for your electronic device to function?

   c. In two brief paragraphs, describe your observations for each half cell during the course of the reaction. Explain your observations chemically.

12

Exploring the Gas Laws

OBJECTIVES

- Quantitatively determine the effect of pressure on the volume of a gas.
- Vaporize a sample of a volatile liquid and determine its mass.
- Use the mass values and the ideal gas equation to determine the molar mass of the volatile liquid.

EQUIPMENT/MATERIALS

Pressure/volume apparatus, meterstick or ruler, water, 400-mL beaker, 125-mL Erlenmeyer flask, Bunsen burner, unknown volatile liquid, aluminum foil, 100-mL graduated cylinder, pin or needle, ring stand, clamps.

INTRODUCTION

Whether you optimistically describe the glass as half full or pessimistically describe it as half empty, you are actually wrong on both counts. Even when it contains no liquid, the glass is completely full—full of air that is. We tend to forget about the sea of air that surrounds us. Gases tend to be invisible and not easily detected. However, when you think about the force that can be exerted by gases (e.g., the wind or the movement of a piston in an engine), it becomes clear that the gas phase is a very interesting state of matter.

Of the three common states of matter (solid, liquid, and gas), gases are the most affected by their physical environment. Solids tend to maintain their shape and volume with only minor changes resulting from alterations in environment (temperature, pressure, type of container, etc.). Liquids assume the shape of their container but tend to maintain their volume. While thermal expansion and contraction due to changes in temperature are noticeable in the case of both solids and liquids, it is generally a rather minor factor. Furthermore, pressure changes do not appreciably affect matter in either of these two states. Gases, on the other hand, are at the mercy of their environment. A given amount of gas, placed in a truly empty (i.e., evacuated) container of any shape, will immediately expand to fill the container completely and uniformly, and the volume of a gas is very sensitive to the pressure and temperature applied to it.

Work by Boyle, Charles, Gay-Lussac, Avogadro, and others studying the behavior of gases in the seventeenth and eighteenth centuries led to numerous theories about matter on the microscopic scale. Unlike liquids and solids, the physical properties of gases tend to be determined more by their physical state than their molecular composition, so these early scientists were able to understand much about matter on the molecular scale from studying macroscopic behavior of gases. In this lab, you will have the opportunity to see and study these relationships firsthand.

You will be measuring the volume of a gas at different pressures (Boyle's law). You will also determine the molar mass of an unknown volatile liquid using the ideal gas law. For the Boyle's law experiment, you will be using an apparatus like the one shown in Figure 12.1. As you can see, one end is closed and the other is open to the atmosphere. The volume V of the gas trapped in the closed end can be determined using the formula for the volume of a cylinder, $V = \pi r^2 h$, where r is the radius of the cylinder and h is its height. As you add water to the open end, the pressure exerted by the water increases and the volume of the gas on the closed side will decrease. You will be measuring the change in volume ΔV at different pressures caused by adding water:

$$\Delta V = V_f - V_i$$

Because the radius r of our cylinder remains constant,

$$\Delta V = \pi r^2 h_f - \pi r^2 h_i = \pi r^2 (h_f - h_i) = \pi r^2 \Delta h$$

$$\Delta V \propto \Delta h$$

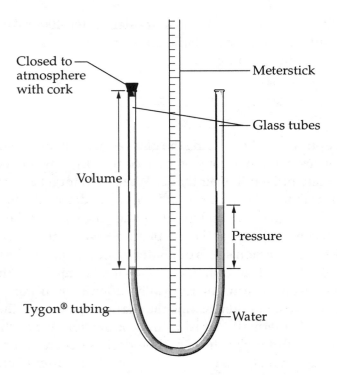

Figure 12.1   Pressure–volume apparatus.

any change in volume is proportional to a difference in h, the length of the column of gas. Therefore, we need only worry about Δh to determine the effect of pressure on volume; there is no need to find the actual volume of the gas by multiplying Δh by the constant factor πr^2.

One can measure pressure in units of mm of Hg, but in order to make the experiment safer and easier, you will be determining pressure in units of cm of H_2O. How do we relate cm of H_2O to the pressure units you are more accustomed to seeing (torr, atm)? Consider the significance of our pressure measurement. Suppose that the water levels were exactly equal on both sides of the tube as is shown in Figure 12.2a. The pressure difference would then be zero, meaning that the pressure exerted by the gas in the closed end is equal to the already-present atmospheric pressure exerted on the open end. Now suppose the height of the column of water on the side that is open to the atmosphere is 10 cm higher than that of the closed end. The pressure inside the closed end would equal the atmospheric pressure added to the pressure exerted by a 10-cm (or 0.10-m) column of water. See Figure 12.2b.

A barometer reads the atmospheric pressure in terms of the height of a mercury column that can be supported. At sea level this is usually about 760 mm Hg or 0.76 m Hg. At a higher altitude, say in Denver, atmospheric pressure is only about 630 mm Hg, or 0.63 m Hg. Mercury is 13.6 times denser than water, so the sea level barometric pressure, in terms of the height of a column of water it would support, would be

$$0.76 \text{ m Hg} \times \frac{13.6 \text{ g / mL } H_2O}{1.0 \text{ g / mL Hg}} = 10.3 \text{ m } H_2O \text{ (1003 cm)}$$

Thus, the pressure that supports a 760-mm column of mercury would support a column of water about three stories high! If the atmospheric pressure was 760 mm Hg during your

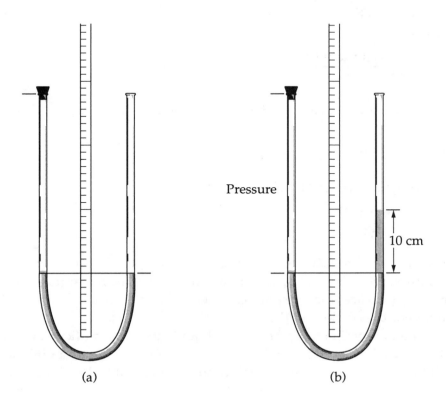

Figure 12.2   (a) Water levels are equal, so $P_{\text{closed end}} = P_{\text{atm}}$. (b) $P_{\text{closed end}} = P_{\text{atm}} + P_{\text{water}}$.

laboratory period, and you observed that the water in the open tube was 10 cm higher than the water in the closed tube of your experimental apparatus, then the pressure inside the closed tube would be

$$1033 \text{ cm H}_2\text{O} + 10 \text{ cm H}_2\text{O} = 1043 \text{ cm H}_2\text{O}$$

Converting this back to mm of Hg, you can use this equation:

$$P = \frac{\text{Height of column of water (cm)}}{1.36 \text{ cm H}_2\text{O} / \text{mm Hg}} = \frac{1043 \text{ cm}}{1.36 \text{ cm H}_2\text{O} / \text{mm Hg}} = 767 \text{ mm Hg}$$

You must remember to add in the atmospheric pressure when determining the pressure of the trapped column of air.

For the second part of the experiment, you will use a version of the ideal gas law: $PV = nRT$. Early scientists learned a great deal about matter on the atomic/molecular level through the study of gas behavior. This is partly because a sample of one gas at a given set of conditions will demonstrate physical properties similar to a sample of another gas at the same set of conditions. The most valuable aspect, with respect to these early scientists, is that the number of molecules or atoms in a sample of gas is proportional to its volume (holding pressure and temperature constant).

If we rearrange the ideal gas equation $PV = nRT$ to solve for n, the number of moles, we obtain

$$n = \frac{PV}{RT}$$

Now let's assign the molar mass of our gas the variable M. The number of moles can then be defined as

$$n = \frac{m(\text{grams})}{M(\text{grams} / \text{mol})}$$

where m = mass of the sample. Substituting for n in the expression above gives

$$\frac{m}{M} = \frac{PV}{RT}$$

Rearranging to solve for M yields

$$M = \frac{mRT}{PV}$$

Thus, we can determine the molar mass M of a sample of gas if we know its volume, temperature, pressure, and mass. You will determine the molar mass of one of several unknown substances that are liquids at room temperature but will vaporize at reasonably low temperatures (below 100°C).

In this experiment, an amount of liquid more than sufficient to fill a flask when vaporized will be placed in a flask covered with foil containing a pinhole. The sample is heated in a boiling water bath, which vaporizes the liquid. The liquid vapor will drive out the air present in the flask and fill the flask with vapor. (Excess vapor will also exit the flask.) When you cool it, the vapor will condense back into a liquid and air will refill the

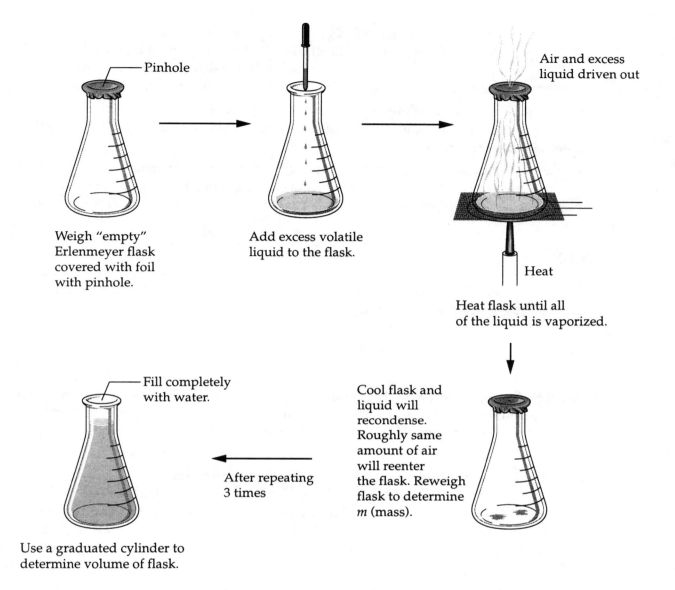

Figure 12.3   A schematic representation of molar mass determination.

Pinhole

Weigh "empty" Erlenmeyer flask covered with foil with pinhole.

Add excess volatile liquid to the flask.

Air and excess liquid driven out

Heat

Heat flask until all of the liquid is vaporized.

Cool flask and liquid will recondense. Roughly same amount of air will reenter the flask. Reweigh flask to determine m (mass).

After repeating 3 times

Fill completely with water.

Use a graduated cylinder to determine volume of flask.

flask. Assuming the liquid in the flask has a negligible volume compared to the flask, the difference in mass between the empty flask and the flask containing the post-vaporized sample will give you the mass of the sample. If you know the pressure, the temperature, and the volume, you can then calculate the molar mass of the sample. A schematic representation of the course of events is shown in Figure 12.3.

PRE-LAB EXERCISES

1. Calculate the pressure exerted in the closed end of an apparatus similar to that depicted in Figure 12.1 if the atmospheric pressure is 720 torr and the height of the water column is 22.3 cm.

2. A flask with a mass of 147.21 g is filled with a sample of volatile liquid and heated to 98°C. Upon cooling, the new mass of the flask with the recondensed liquid is found to be 147.68 g. The volume of the flask is determined to be 135.6 mL, and the atmospheric pressure is 743.5 torr. Calculate the molar mass of the volatile liquid.

3. If there was an error in the pressure measurement in problem 2, such that the atmospheric pressure was in reality *lower* than that measured, how would this affect the value determined for the molar mass?

4. Prepare your notebook for this experiment. You will need to prepare data tables to collect volume/pressure (or rather air column height/water column height) data in the first part, as well as the masses, volume, and temperature measurements made in the second part.

PROCEDURE

Record in your notebook the atmospheric pressure in the lab on the day of the experiment. Your instructor will either provide this information or show you how to obtain it.

1. Effect of Pressure on a Gas Volume

To investigate this effect, you will use an apparatus consisting of a meterstick with two 3-foot sections of 10-mm glass tubing bound to it with rubber bands. These sections of glass tubing are connected together at the bottom with a piece of Tygon® tubing. A small cork is supplied to close the upper end of one of the tubes; the end of the other tube remains open (Figure 12.1).

With the cork removed from the glass tube, add just enough water so that the loop of Tygon® tubing is filled and the water level is at about the 1-cm mark on the meterstick. The level should be equal on both sides. Put the cork in the top of the glass tube. (If the cork is soaked in water for a minute or so before you insert it, your seal will be more airtight.) The space between the corked end of the closed tube and the top of the water level in that tube contains the gas (air) whose volume changes will be measured.

Record the exact position of the water level against the meterstick on the side with the closed tube. Record the exact position of the bottom of the cork. (Subtracting the one value from the other gives you the height of the column of air trapped in the tube.)

Record the exact position of the water level against the meterstick on the side with the open tube. The difference in height between the two liquid levels allows us to calculate pressure being exerted on the gas in the closed tube by the column of water.

Now add some water to the open tube, enough to raise the level in that tube by about 8 or 10 cm. Record the level in the open and closed tubes. Repeat this procedure several times until you have filled the open-ended tube, each time carefully reading the liquid levels in both tubes to the nearest 0.1 cm. Record these values.

2. Determination of the Molar Mass of a Volatile Liquid

A diagram of your reaction apparatus is shown in Figure 12.4. Fill a 400-mL beaker about two-thirds full with water. Start heating the water with a Bunsen burner.

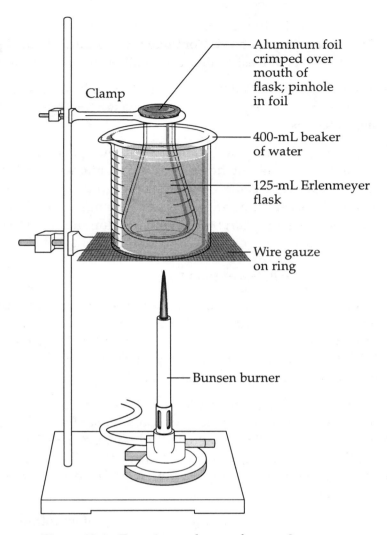

Figure 12.4   Experimental setup for part 2.

While the water is heating, obtain a clean, *dry* 125-mL Erlenmeyer flask and weigh it together with a square of aluminum foil. Record this combined mass in your notebook.

Pour ~2–3 mL of your assigned unknown liquid into the flask.

CAUTION: DO NOT INHALE THE VAPOR FROM THE LIQUID.

Crimp the aluminum foil over the mouth of the flask to form a cap; try to seal it as completely as possible. Fold the foil up so that it doesn't hang down too far on the neck of the flask (~1 cm is far enough). Using a pin, poke a small pinhole in the top of the foil.

Set up the apparatus shown in Figure 12.4. Carefully clamp the flask to the ring stand. (*Note*: If you angle the flask slightly, it is often easier to see when all the liquid has vaporized.) Immerse as much of the flask as possible in the water bath (but don't let it touch bottom). Add additional hot water to the beaker if necessary to get maximal immersion.

Heat the water bath to boiling and measure the temperature of the boiling water. Record this temperature in your notebook. Continue heating until all the liquid inside the flask has vaporized, and heat for at least 1 minute after all the liquid has vaporized.

Carefully remove the flask from the hot water bath. Allow it to cool to room temperature. Make sure it is completely dry and carefully reweigh the flask with the foil. Record this mass in your notebook.

Your instructor may direct you to obtain data for multiple runs. If so, remove the foil cap, discard the condensed liquid in the proper waste container, add 2–3 mL of additional liquid, and repeat the heating-cooling-weighing procedure one or more times. Make sure that you reweigh the flask each time after it cools. If you need to replace your aluminum square, you must reweigh the empty (*dry!*) flask with the aluminum foil to obtain a new initial weight.

Determination of the Volume of the Flask

After completing the procedure above as many times as your instructor directs, you will determine the volume of your flask. (*Hint:* It is not equal to 125 mL.) Discard the condensed liquid from your last run and discard the aluminum foil. Rinse the flask 2–3 times with water. Fill the flask completely, all the way to the rim with water. Tap the flask gently to make sure there are no bubbles clinging to the side.

Carefully pour the water from the flask into a 100-mL graduated cylinder. You will need to do this in two stages, as the graduated cylinder will not be able to hold all the water in the flask. To accomplish this, fill the graduated cylinder up to ~95 mL from the flask. Note (and record) the exact volume, and discard the water. Fill the graduated cylinder with the rest of the water from the flask and note (and record) the exact volume. The sum of these two volumes is the volume of the flask.

Repeat this process two more times. Take an average of your three values for the flask volume.

REPORT

Part 1

1. Prepare a table that has the following headings:
   - Height of Column of Water
   - Pressure of Column of Trapped Air
   - Height of Column of Trapped Air

2. Enter your data for the first and third of these (Height of Column of Water and Height of Column of Trapped Air). Calculate the second, Pressure of the Column of Trapped Air, using the equation $P_{air} = P_{atm} + P_{H_2O}$ where P_{atm} is the atmospheric pressure in the lab (mm Hg) and P_{H_2O} is determined using the equation

$$P_{H_2O} = \frac{\text{Height of column of water (cm)}}{1.36 \text{ cm } H_2O \, / \, \text{mm Hg}}$$

3. Graph the relationship between Pressure of Column of Trapped Air and Height of Column of Trapped Air (which is proportional to the volume). Your instructor might direct you to use a computer graphing program to accomplish this. (Keep in mind that a curvature over a small distance may *appear* linear. Also remember that it is customary to plot the independent variable along the *x*-axis and the dependent variable along the *y*-axis.) Which is the independent variable in this case?

4. Prepare a new table that has the headings

   ■ Pressure of Column of Trapped Air
   ■ Inverse of the Height of Column of Trapped Air $\left(\text{i.e., } \dfrac{1}{\text{Height}} \right)$.

5. Graph the new relationship between Pressure of Column of Trapped Air and 1/Height. How does this graph differ from the previous graph?

Part 2

1. Report your data for the following:

   ■ Mass of volatile liquid in flask after cooling (average if you performed more than one trial)
   ■ Volume of flask
   ■ Atmospheric pressure on the date of the experiment (in atmospheres)

2. Use these values and the formula below to determine the molar mass, using the average volume from your three trials to measure the volume of the flask. (Don't forget to convert your temperature to the absolute scale: $T(K) = T(°C) + 273.15$; $R = 0.0821 \text{ L} \cdot \text{atm} / \text{K} \cdot \text{mol}$.)

$$M = \frac{mRT}{PV}$$

3. In actual fact, the amount of air that reenters the flask is different from that which was present upon initial weighing and was subsequently expelled from the flask. Because the liquid you used was a volatile one (otherwise the experiment won't work), it didn't completely condense. Some of it remained in the vapor phase, even at room temperature. Is the amount of air that reenters greater or less than that which was present upon initial weighing? If you were able to correct for the difference (and you can, using the density of air and the vapor pressure of the volatile liquid), would your final molar mass value be greater or less than the value you determined? Explain your reasoning.

13

Properties of Liquids

OBJECTIVES

- Explore different properties of liquids: surface tension, viscosity, vapor pressure, and capillary action.
- Relate those properties to intermolecular attractive forces in a qualitative fashion.

EQUIPMENT/MATERIALS

Water, detergent solution, propylene glycol, mineral oil, glycerin, hexane, acetone, methanol, watch glass, pieces of wire, paper towels, buret, glass beads, round-bottom flask, stopper, masking tape, ring clamp, ring stand, plastic bag with ice, ruler, stopwatch or watch/clock with a second hand, beakers, flasks.

INTRODUCTION

A liquid can be described by numerous properties, all of which relate to the *intermolecular forces* (IMF) present. Read the section in your textbook that describes IMF. In this experiment, you will explore the following properties:

- Surface tension
- Viscosity
- Vapor pressure
- Capillary action

Surface Tension

The molecules at the surface of a sample of liquid experience a very different environment from the one experienced by those in its center. In the center, the IMF are distributed symmetrically in space among all the molecules, but at the air–liquid interface, the molecules are attracted only by the molecules below and beside them. This causes the surface of a liquid to contract and act like a "skin." Surface tension is the force required to penetrate this skin. The stronger the intermolecular forces present in the liquid, the greater the surface tension. You will explore the surface tension of water in a qualitative fashion.

Viscosity

Viscosity is defined as a resistance to flow. Liquids such as honey or molasses are very viscous, whereas the viscosity of water is very low. When a liquid flows, the molecules slide past each other. This movement is impeded by strong IMF. The viscosity of a liquid decreases with increasing temperature. You may have seen television commercials that describe the thermal breakdown of motor-oil viscosity: as an automobile engine heats up, its motor oil becomes less viscous. It is imperative that the motor oil maintain as much viscosity as possible to lubricate all the necessary engine components. Viscosity failure could mean disaster for the engine.

A simple method to compare the viscosities of different liquids (particularly liquids of high viscosity) is the falling-sphere method. You drop a sphere into a liquid and measure the time required for the sphere to travel a calibrated distance. You will compare the viscosities of propylene glycol, glycerin, and mineral oil.

Vapor Pressure

The vapor pressure of a liquid is defined as the pressure of the volatile gas above a liquid at equilibrium. (See your textbook for a more detailed explanation.) It is dependent on (1) IMF, (2) temperature, and (3) atmospheric pressure.

1. The stronger the IMF are in a given liquid, the lower its volatility. A less volatile liquid will have a lower vapor pressure at a given temperature than a liquid with high volatility.

$$\text{As IMF} \uparrow, P_{vap} \downarrow$$

2. As the temperature of a liquid increases, more of its molecules have enough energy to escape the liquid phase and enter the vapor phase. Thus, its vapor pressure will increase with increasing temperature.

$$\text{As } T \uparrow, P_{vap} \uparrow$$

3. As the atmospheric pressure pushing down on the liquid decreases, liquid molecules can escape into the gas phase with less energy. Thus, at lower pressure, the vapor pressure increases.

$$\text{As } P_{atm} \downarrow, P_{vap} \uparrow$$

When the vapor pressure of a liquid equals the atmospheric pressure, the liquid is said to be "boiling." The temperature at which this occurs is called the *boiling point* of the liquid, and when a liquid is at its boiling point, its molecules have enough energy to enter the gas phase anywhere in the bulk liquid, not just at the surface. This is why bubbles of vapor form in a boiling liquid. Because the atmospheric pressure at higher altitude (such as in the mountains of Colorado or Tibet) is lower than the atmospheric pressure at sea level, the boiling point of a given liquid is also lower. Indeed, the atmospheric pressure is so low on the top of Mount Everest that bacteria can survive in boiling water. In this laboratory, you will explore boiling liquids at reduced pressure.

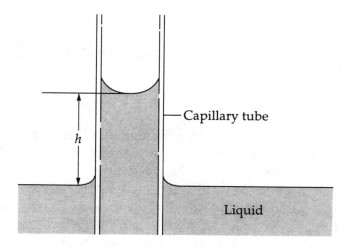

Figure 13.1 Capillary action.

Capillary Action

Intermolecular forces can occur between like molecules (cohesive) or between a substance and a surface (adhesive). It is the adhesive forces that allow you to use a paper towel to mop up a puddle of water. The water molecules are attracted to the cellulose molecules (which are rich in –OH groups for hydrogen-bonding) in the paper. There is little adhesive force between water molecules and a surface such as plastic wrap (which is held together pre-dominantly by nonpolar dispersion forces), which is why plastic wrap is so unsatisfactory at soaking up spills and so useful at excluding water.

In previous laboratory activities, you have noticed the meniscus, or curved surface, of liquids in glassware. The liquid adheres to the glass molecules and crawls up the side of the vessel because the IMF between the glass and the liquid are strong enough to overcome gravity—up to a point. Many liquids will rise in a small-diameter tube, called a capillary tube, to a distance, h, that is related to the strength of the IMF between the liquid and the glass. (See Figure 13.1.) This is called *capillary action* and is responsible for the upward movement of water and nutrients in plants and the soaking of liquid into a sponge.

PRE-LAB EXERCISE

1. Make an outline of the procedure. Include any Further Investigation parts (see below) that your instructor assigns.

2. Prepare your notebook for this experiment. You will need to prepare the following data tables. In addition you will make and record qualitative observations for the sections on surface tension and vapor pressure, as well as any Further Investigation sections you are assigned.

   - *Viscosity*: This table should have headings for liquid identity and time for glass bead to fall. (You will do four or five trials per liquid.)

   - *Capillary action*: This table should have headings for liquid identity and height of capillary column. If you do the Further Investigation, you'll need a second table with headings for temperature and height of capillary column.

PROCEDURE

At the end of each section, you will see a section labeled Further Investigation. Your instructor will inform you which, if any, of these to perform.

1. Surface Tension

Obtain a large watch glass and wash it thoroughly. Fill it with water. Obtain one of the small pieces of iron wire provided. Tear off a small piece of paper towel that is slightly larger than your piece of wire. Place the wire on the paper and use your finger or a spatula to carefully float the paper containing the wire on the water surface. As the paper towel gets soaked with water, it should sink, leaving the wire floating on the surface of the water. (If you have difficulty, you may wish to rewash your watch glass. Also make sure that the wire is clean and dry, and avoid handling it too much with your greasy fingers!)

Observe the floating wire carefully from the side as well as the top. Add a drop of the detergent solution provided to the water in the watch glass. Try to disrupt the surface as little as possible and add the drop as far away from the floating wire as possible. What happens?

Further Investigation: With practice, you can float a paper clip, a thumbtack, or a rubber band. Experiment with different concentrations of detergent solution. Try adding something else (e.g., acetone, methanol, sugar solution) to the water. More than one drop may be required to produce an effect. You may also wish to explore the effect of temperature on surface tension. As before, make sure your watch glass is scrupulously cleaned between experiments.

2. Viscosity

You will measure the time required for a sphere to move through a prescribed distance in each of the liquids listed below to compare their relative viscosities.

Obtain a buret. Use a ruler to measure the distance between the 10-mL mark and the 40-mL mark on the buret. Add approximately 45 mL of the liquid to the buret. Carefully drop one of your glass beads into the buret and measure the time (in seconds) required for it to fall the prescribed distance between the 10-mL and 40-mL marks. You may need to work with a partner to do this effectively. Repeat the measurement for three additional beads. Measure for a fifth bead if the first four measurements are not close to each other. After following the rinsing and drying instructions below, repeat this process for the other two liquids.

Pour out the liquid into the appropriate container. These liquids will be recycled, so be sure that the liquid you are pouring back into the container matches the container label. If there is any doubt in your mind, use the provided waste container instead. Avoid dumping your beads out into the recycling container. Rinse the buret thoroughly, using the rinsing agent suggested in the table below. Rinse and dry your beads carefully. Invert your buret and allow it to dry between measurements. While it is drying, you can perform the other parts of this laboratory experiment.

Liquid	Rinsing Agent
Propylene glycol	Water
Mineral oil	Detergent/hot water
Glycerin	Water

Further Investigation (choose one of the following):

a. Fill a buret with 20 mL of mineral oil and 20 mL of propylene glycol. Allow the layers to separate and note which layer is above and which is below. Drop a glass bead through the liquids. Observe what happens. Consider the properties of density versus viscosity with respect to intermolecular forces and what is happening at the molecular level when the sphere moves through the liquid. **Do *not* pour this mixture back into the "recycled" liquid containers. It must be disposed of in the labeled waste container.**

b. Choose one of the liquids provided and determine how its viscosity changes with changing temperature. Heat it carefully as all of these liquids are flammable, if not volatile.

3. Vapor Pressure

Assemble the apparatus shown in Figure 13.2. Add approximately 50 mL of tap water to a 500-mL round-bottom flask. Heat the flask until the water boils; allow it to boil for at least 2–3 minutes. Remove the heat and allow the water to stop boiling. Firmly insert a stopper into the neck of the flask. Use a paper towel to hold the neck of the flask, as it will be hot. Wrap masking tape around the stopper to be sure that it is held in place. **(CAUTION!)** Again using your paper towel to hold the flask, carefully invert it in the ring (see Figure 13.2). Place a plastic baggie full of ice directly on top of the inverted flask. Observe and record what happens.

CAUTION: BE SURE YOUR STOPPER IS SECURE IN THE NECK OF THE FLASK TO AVOID BEING SCALDED BY HOT WATER.

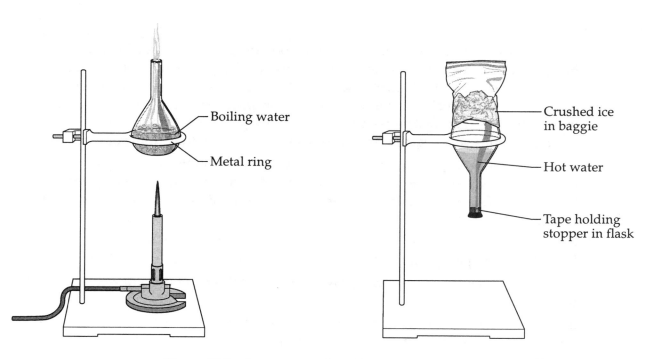

Figure 13.2   Apparatus to investigate vapor pressure.

Periodically remove and replace the baggie. Observe and record what happens. After the flask is cool and the water is no longer boiling, remove the stopper, observing what happens when you do so. Record your observations.

Further Investigation: Obtain one of the special taped test tubes with a side arm and stopper. Clamp the test tube firmly. Add 4–5 mL of acetone to the test tube and insert the stopper. Connect the hose to the vacuum and slowly open the vacuum line. Allow the test tube to remain under vacuum for several minutes while you observe what happens.

4. Capillary Action

Label five small test tubes A–E. Place about 3 mL of each of the five liquids listed below in the five test tubes (one liquid per tube). Place a capillary tube in each of the test tubes. Use a ruler to measure the height, h, of the liquid in the capillary tube.

Water	Acetone **(CAUTION!)**
Hexane **(CAUTION!)**	Propylene glycol **(CAUTION!)**
Methanol **(CAUTION!)**	

CAUTION: FLAMMABLE. KEEP AWAY FROM FLAMES.

Further Investigation: Investigate the effect of changing temperature on the capillary action of *water only* (do not heat the flammable liquids). Compare water that is cold (from an ice bath), room temperature (~20°C), warm (~ 50°C), and hot (almost boiling). As best you can, measure the heights of the capillary columns.

REPORT

1. Write a description of the procedure and your observations for each of the sections in the procedure. Discuss your results and any conclusions that you can draw from them within the context of the effects of different kinds of IMF. Include your data tables where appropriate. Some additional guidelines for each section are given below. For any of the Further Investigation sections that you completed, include a paragraph that outlines your experiment, the results you obtained, and the conclusions you can draw from your results.

2. In the experiment exploring surface tension, what effect did the detergent have? Detergents are also known as *surfactants*. Surfactant is a term that comes from squeezing together "surface acting agents." Comment on the suitability of this name given your experimental results.

3. In the viscosity section, use your data to rank the viscosity of the three liquids you tested. If available, compare your results to literature values for the viscosity of these liquids.

4. In the vapor pressure experiment, provide an explanation for how you were able to boil water with ice!

5. In the capillary action section, rank your liquids with respect to capillary action (based on the h you measured), and comment on the relationship between these rankings and the IMF present.

14

Structure–Solubility Relationships in Molecules

OBJECTIVES

- Practice structural analysis of molecules.
- Practice assigning polarity to appropriate molecular areas.
- Examine the relationship between solubility, polarity, and structure.
- Determine whether common molecules are soluble in water (a polar solvent) or cyclohexane (a nonpolar solvent).
- Explore relationship between solubility and temperature.

Review

- Solubility.

EQUIPMENT/MATERIALS

Test-tube racks, test tubes (two for each solute tested), eyedroppers, beakers, molecular model kits (optional), water, cyclohexane, ethanol, acetic acid, citric acid, benzoic acid, glucose, 1-octanol, cholesterol, ascorbic acid, tocopherol.

INTRODUCTION

Every time you stir sugar into your coffee or tea you are making a solution. Indeed, the coffee and tea were solutions before you even added the sugar. A *solution* is broadly defined as a two-part system consisting of a solute (or solutes) and a solvent. The *solvent* is the substance in which the solute(s) is (are) dissolved. It can also be described as the substance present in greater quantity. The *solute* is what is being dissolved, and it is the substance present in lesser quantity. When water is the solvent, a solution is described as *aqueous*. Most drinks such as coffee, tea, soda, and juice are examples of aqueous solutions. Physiological solutions such as blood plasma, urine, sweat, and tears are also aqueous solutions. *Nonaqueous* solutions are less common in daily life, but examples include mineral spirits, paint thinner, and gasoline. In all these examples, the solvent is something other than water.

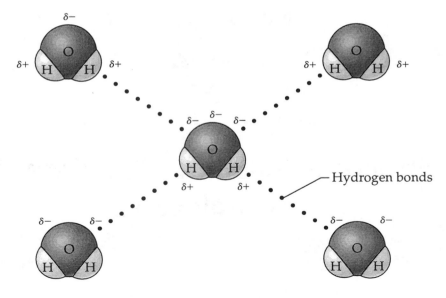

Figure 14.1   Hydrogen bonds among water molecules.

In the solution state, substances are intermingling at the molecular level. The ability of a compound to form a solution by interaction with other molecules at the molecular level is called *solubility* and is stated neatly in the rule "like dissolves like." Substances that are polar typically dissolve and are dissolved in other polar substances. Conversely, substances that are nonpolar will dissolve and are dissolved in other nonpolar substances. The mixing at the molecular level occurs through the same intermolecular attractive forces that govern the behavior of liquids and solids.

Polar molecules interact with each other through similar attractions. In Figure 14.1 you can see the hydrogen bonds that form among different water molecules. These are not bonds in the true sense because there is no sharing of electrons. They are based on attractions between slightly negative areas of one molecule (oxygen) and slightly positive areas of another (hydrogens). In order for a substance to dissolve in water, it has to squeeze between water molecules and form attractions similar to hydrogen bonds. In order to do this, it must have a slightly negative and/or a slightly positive area to which water molecules can be attracted; in other words, it must be polar. Ionic species and polar molecules are able to form attractions with water molecules that result in the formation of a solution, as shown in Figure 14.2. Molecules that are polar and can interact with water are called *hydrophilic* or "water-loving." Polar molecules are easily transported in the blood and other body fluids because of their solubility.

Nonpolar molecules are not attracted to water and cannot intermingle with water at the molecular level. These molecules will be squeezed out of the water and will congregate as separate phases, as in a lava lamp or oil-and-vinegar salad dressing. Nonpolar molecules are called *hydrophobic* or "water-fearing." Because they cannot dissolve in water, they must have help when being transported through the blood or aqueous body fluids. Molecules that are *amphipathic* provide this assistance. Amphipathic molecules have a nonpolar "tail" and a polar "head." (See Figure 14.3a.) They are very important as structural components in cell membranes and as transport molecules for nonpolar substances in the blood. In aqueous solutions, such as fluids contained within the body, amphipathic molecules form *micelles*. The polar areas face out into the water, and the nonpolar areas associate inside the micelle away from the

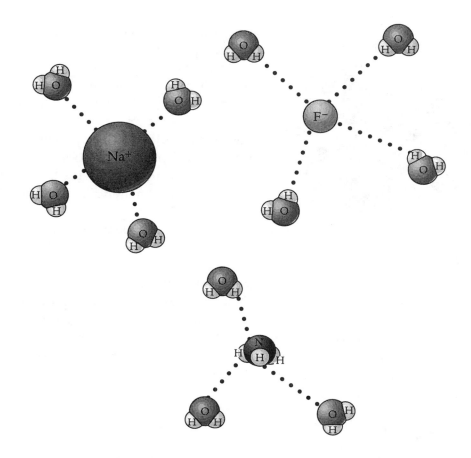

Figure 14.2   Ion–dipole and hydrogen-bonding interactions between water and polar substances.

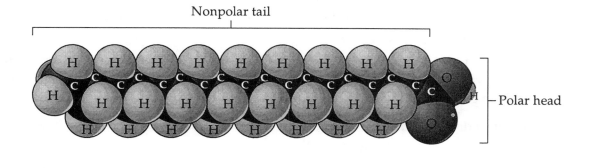

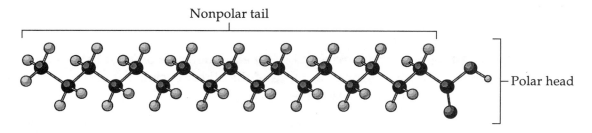

Figure 14.3   (a) Stearic acid, an example of an amphipathic molecule.

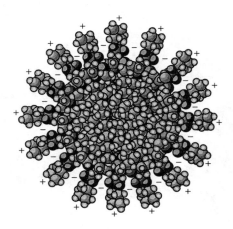

Figure 14.3   (b) Cross-section of a micelle in an aqueous medium.

water. Think of micelles as spheres with a nonpolar interior and a polar skin. See Figure 14.3b.

In the body, solubility refers to the ability to dissolve either in blood or other watery liquids or in fats. The transport and storage of water-soluble molecules is different from that of fat-soluble molecules. The way in which these molecules interact with cell membranes and proteins is different. Interactions with water also determine the shape that a molecule has in physiological solutions and, ultimately, its function in the body.

In this laboratory unit, you will analyze the structures of some common molecules to predict whether they will dissolve in water, which is a polar solvent, or in the nonpolar solvent, cyclohexane. The solutes range from relatively small organic molecules like ethanol to relatively large biomolecules like cholesterol and are shown in Figure 14.4. The structures of the two solvents, water and cyclohexane, are also shown in Figure 14.4. You want to determine what kinds of interactions could occur between each solvent and each biological molecule by examining their structures. You will test your analysis by observing the actual solubilities of specified molecules in both solvents and by comparing your results to published solubility data.

PRE-LAB EXERCISE

1. Prepare a table in which you list the following information:

   ■ The name, formula, and molar mass for each substance used in the experiment
   ■ Its approximate three-dimensional molecular structure (draw it, if that's easier)
   ■ The types of molecular interactions (polar or nonpolar) that it is likely to exhibit
   ■ Its predicted solubility

   Line-and-stick drawings of each of the molecules are shown in Figure 14.4. Your instructor may suggest that you build models of the molecules or direct you in using a computer visualization program or Web site to help you visualize the structures and predict the polarity.

2. Prepare a data table for this experiment wherein you can list each substance and its relative solubility in hexane and water.

PROCEDURE

1. Solubility

Obtain a test-tube rack and two test tubes for each molecule on your list. Make sure to label the test tubes and designate one "water" and the other "cyclohexane." (Your instructor may ask you to use a well plate instead of test tubes.)

Place a small amount of each solute in the test tube with its name. For liquids, two small drops should be enough. For solids, a single small crystal or a tiny bit of powder will be enough to see dissolution.

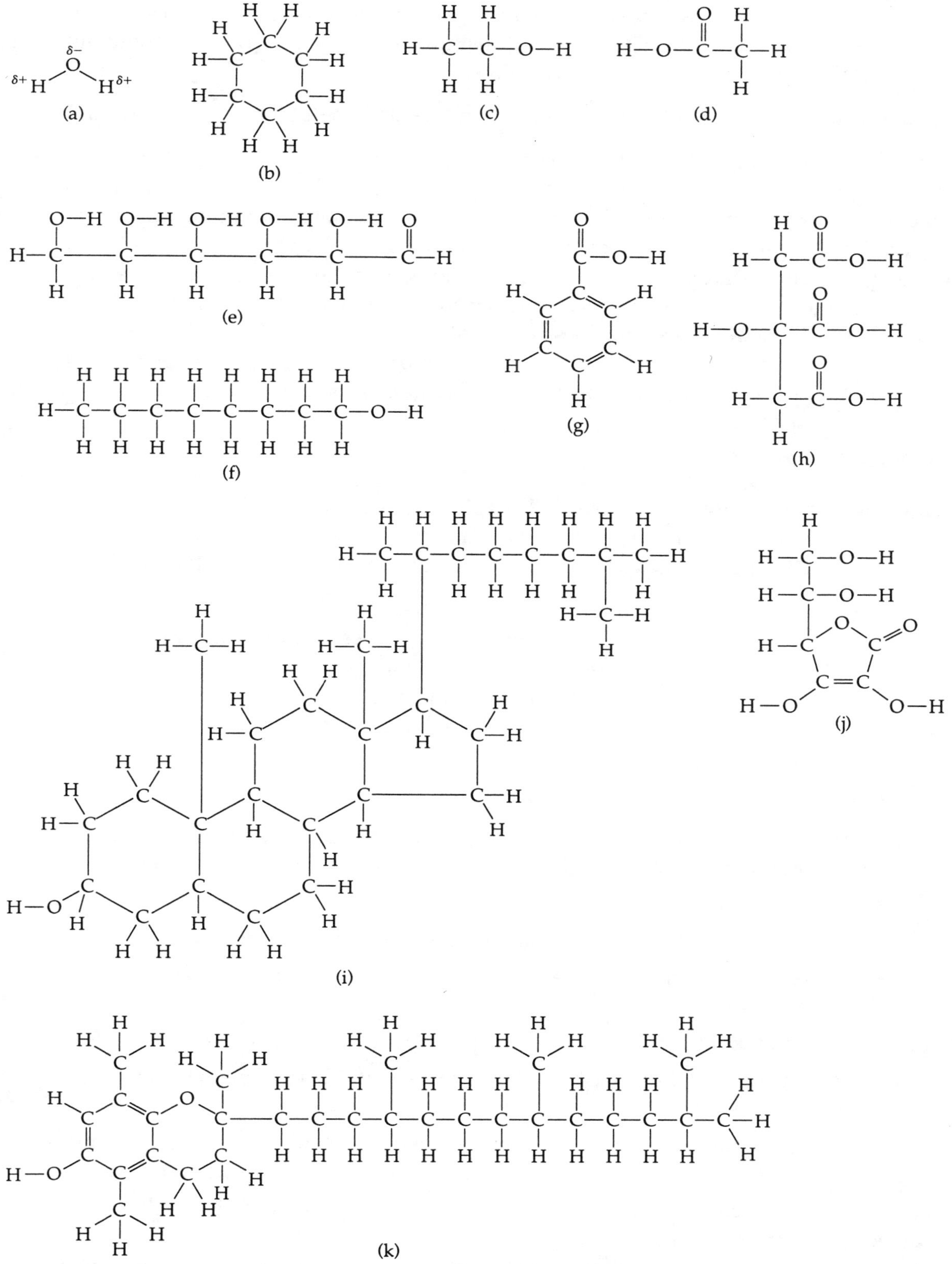

Figure 14.4   Molecular structures.
(a) Water, (b) Cyclohexane, (c) Ethanol, (d) Acetic acid, (e) Glucose, (f) 1-Octonol, (g) Benzoic acid, (h) Citric acid, (i) Cholesterol, (j) Ascorbic acid (vitamin C), (k) Tocopherol (vitamin E)

Add one drop of water to each test tube and shake. Note whether the substance dissolves. If it does not, add another drop of water and shake again. Record the observed solubility in your data table. Record any special observations.

Repeat this process using cyclohexane. Record the observed solubility in your data table.

How do you know whether a solid or liquid has dissolved? For solids it's easy: as the substance forms attractions with water molecules, it becomes incorporated into the molecular matrix of the water and disappears. Liquids that are soluble may initially form little "waves" as they are added to the solvent. These look very much like the heat waves that rise from a hot road in the summer. As the liquid molecules are incorporated into the solvent, these waves will disappear. Sometimes when a liquid solute is added to a liquid solvent, the mixture becomes cloudy. This is a sign that the solute does not dissolve in the solvent. If a solid or liquid is non-polar, it may either float to the top of the water surface or fall to the bottom, depending on its density. The separation between polar and nonpolar areas is usually fairly distinct.

2. Temperature Effects on Solubility

Weigh out 0.5 g of benzoic acid. Place the benzoic acid in a clean 25-mL Erlenmeyer flask. Measure out 12 mL of water in a graduated cylinder and add about 5 mL to the benzoic acid. Swirl the flask gently and note whether or not the benzoic acid dissolves. Heat the flask on a hot plate to a gentle boil.

Continue heating while you add water in ~0.5-mL increments until no more solid appears to dissolve. (You shouldn't need more than 12 mL total.)

Cool the flask in an ice bath. Note and record what happens.

REPORT

1. Explain the solubility results you obtained in terms of the structure of each molecule. Examine your results for relationships between the following:
   - Solubility in each type of solvent and the presence of polar groups
   - Solubility in each type of solvent and the presence of nonpolar groups
   - Solubility in water and the relative number of polar and nonpolar areas of the molecule
2. Compare your experimental results to your pre-laboratory predictions.
3. Were there any molecules that dissolved in both solvents? If so, explain the observation in terms of the molecule's structure and polarity.
4. Can you use your results to determine whether there is a relationship among the number of –OH groups in a molecule, the number of carbon atoms, and its solubility in water?
5. Some vitamins, such as vitamin A and vitamin E, are nonpolar and dissolve in fat. These can be toxic if taken in large doses. Other vitamins, such as vitamin C, are polar and dissolve in water. These vitamins are typically not toxic if taken in large doses. Explain this difference in toxicity.
6. Describe the solubility of benzoic acid in water with respect to temperature.

15

Salt Solutions: Preparation, Density, and Concentration Relationships

OBJECTIVES

- Understand the meaning and use of various concentration terms.
- Learn how to prepare solutions of specified concentration from a solid and using serial dilution of a stock solution.
- Examine the relationship between density and different concentrations of solutions.
- Construct a standard graph of concentration–density relationships for a set of standard solutions.
- Determine the concentration of an unknown salt solution by measuring its density.

Review

- Pipetting techniques.
- Use of balances and mass measurements.
- Use of volumetric ware and volume measurements.
- Significant figures in measurements and calculations.
- Graphing techniques.

EQUIPMENT/MATERIALS

Four 100-mL volumetric flasks, beakers, funnel, stirring rod, 10-mL graduated cylinder, 50-mL graduated cylinder, water, 10-mL pipets, NaCl, weighing boat, hydrometer to measure specific gravity, unknown salt solutions.

INTRODUCTION

Solutions and Concentration

Solutions and their components were introduced in the previous laboratory unit. We discussed how solutions formed, and explored the solubility of polar and nonpolar substances. In this experiment, we will explore the quantitative aspects of solutions. You will learn good technique in solution preparation as well as develop an understanding of concentration.

Salts typically dissociate when they dissolve in water. Dissociation is the separation into individual ions. For example, when the salt, NaCl, dissolves in water, it dissociates into sodium ions and chloride ions, as shown below.

$$NaCl(s) \longrightarrow Na^+(aq) + Cl^-(aq)$$

The amount of solute per unit volume or mass of solution is called the *concentration*. Concentration information is quantitative and tells you not just what solute is present but exactly how much of it is present. Solution concentrations are reported using a variety of terms. The units in which concentration is reported depend on which properties of the solution are of interest. Some common concentration terms are defined in Table 15.1.

Table 15.1   Common Concentration Units Used in Chemistry

Concentration Term	Use
Molarity (M)	Tells you the number of moles of solute per liter of solution. Useful when reporting numbers of chemical species (molecules) in a solution; used for stoichiometric calculations. $$Molarity = \frac{Moles\ of\ solute}{Liters\ of\ solution}$$
Molality (m)	Tells you the number of moles of solute in a kilogram of solvent. Used for precise physical measurements. $$Molality = \frac{Moles\ of\ solute}{Kilograms\ of\ solvent}$$
Weight/volume %	Tells you the number of grams of solute in 100 mL of solution. Commonly used to describe a solid solute in a liquid solvent. $$Weight\ /\ Volume\ \% = \frac{Weight\ of\ substance\ (in\ grams)}{Volume\ of\ total\ solution\ (in\ milliliters)} \times 100$$
Weight %	Tells you the number of grams of solute in 100 grams of solution. Commonly used to describe a solid solute in a solid solvent. $$Weight\ \% = \frac{Mass\ of\ substance}{Mass\ of\ total\ solution} \times 100$$
Volume %	Tells you the number of milliliters of solute in 100 mL of solution. Used to describe a liquid solute in a liquid solvent. $$Volume\ \% = \frac{Volume\ of\ substance}{Volume\ of\ total\ solution} \times 100$$
ppm/ppb	Tells you amounts of solute per million (ppm) or billion (ppb) parts of solution. Used to report minute amounts of solute, such as levels of pollutants or toxic substances in air and water.

Figure 15.1   Measure the desired mass of solute.

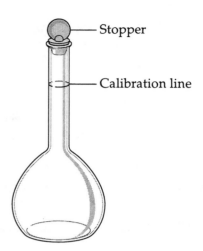

Figure 15.2   A volumetric flask with a calibration line on the neck.

Solutions are usually prepared in one of two ways. In the first method, a specified amount of solute is measured (usually by weighing) and dissolved in an initial volume of solvent. (See Figure 15.1.) Additional solvent is then added to provide the desired final volume. This method requires the use of a *volumetric flask* that has only one calibration line on the neck. This calibration line is the level to which solvent is added (see Figure 15.2). Figures 15.3 and 15.4 outline the steps in this type of preparation.

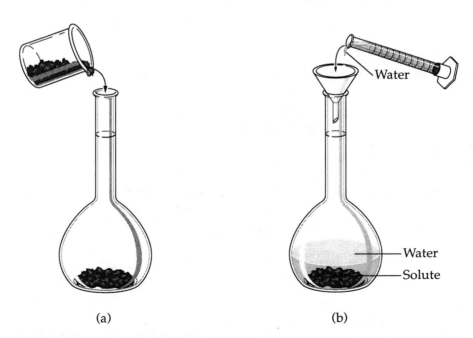

(a)   (b)

Figure 15.3   (a) Transfer the measured mass of solute to a volumetric flask.
                (b) Add an initial volume of water (the solvent) to dissolve the solute.
                    Do not add water to the calibration line.

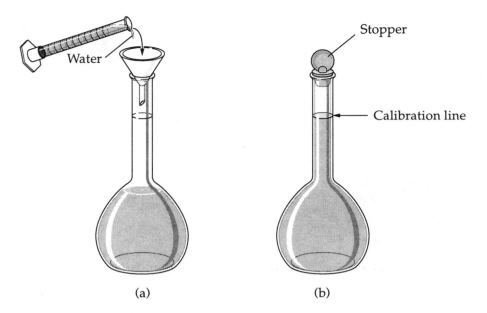

Figure 15.4   (a) Add water (the solvent) to the calibration line.
(b) Stopper the flask and invert it to mix the contents thoroughly.

The second common method of solution preparation is the dilution of a more concentrated solution, known as a *stock solution*, with solvent to produce a less concentrated solution of desired concentration. A *volumetric pipet* is used to transfer a specified volume of the more concentrated stock solution to a volumetric flask. The solution is then brought to the desired volume by adding solvent. Figures 15.5 and 15.6 outline the steps in this second type of preparation. The following is a useful equation for the dilution of a stock solution:

$$C_{stock}V_{stock} = C_{dilute}V_{dilute}$$

where C represents the concentration and V, the volume. The concentration times the volume represents the number of moles. Thus, the number of moles in the dilute sample is the same as is in the sample of concentrated stock solution. They are just spread out over more solvent to give a more dilute concentration.

Density of Solutions

The density of an aqueous solution is directly related to the amount of solid solute dissolved. Recall that the density of a substance is defined as the ratio of the mass of a sample of the substance to the volume the sample occupies:

$$\text{Density} = \frac{\text{Mass}}{\text{Unit volume}}$$

For the solutions in this laboratory unit, the mass is measured in grams (g) and the volume is measured in milliliters (mL).

Suppose you make two solutions. The first has 1.0 g of NaCl dissolved in 100 mL of solution. The second has 3.5 g of NaCl dissolved in 100 mL of solution. Because more solute is dissolved in the second solution and the two solutions have equal volumes, the second will

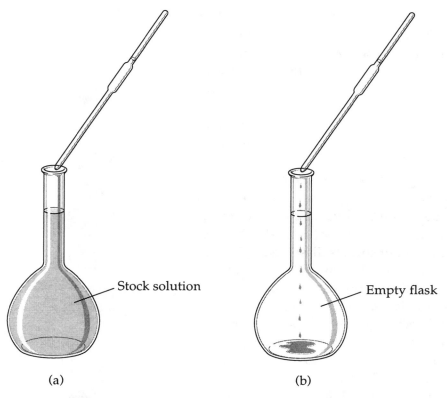

Figure 15.5 (a) Pipet the desired volume of the stock or starting solution.
(b) Transfer the desired volume of the stock or starting solution to an empty volumetric flask.

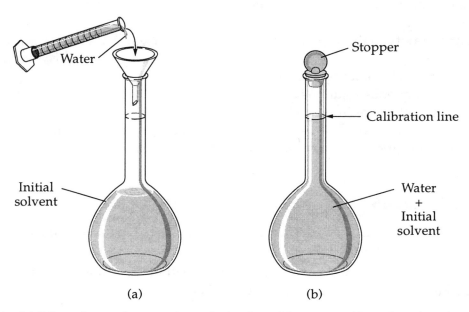

Figure 15.6 (a) Dilute the stock or starting solution by adding water (the solvent) to the calibration line.
(b) Stopper the flask and invert it to mix the contents thoroughly.

have a greater density. Mathematically, when comparing the density of the two solutions, the numerator increases and the denominator stays the same. The effect, then, is that the density of a solution increases as more solute is added to the same volume of the solution.

Beer and wine brewers use density to monitor the progress of fermentation in their products. Microorganisms in wine and beer change sugars to alcohol and carbon dioxide with a corresponding decrease in the density of the beverage. A physiological solution such as urine is an aqueous solution that has a specified amount of solutes dissolved in it under normal (i.e., healthy) circumstances. Deviations from this normal range can indicate health problems.

Specific gravity can be measured with a device called a hydrometer (see Figure 15.7). Hydrometers are weighted, bulb-shaped instruments that float in a solution. When the hydrometer is placed in a solution, it displaces an amount of solution equal to the hydrometer's weight. The more dense a solution, the higher the hydrometer floats because less liquid needs to be displaced in order to equal the hydrometer's weight. The calibration scale is on the hydrometer neck and increases down the stem. Hydrometers allow you to report specific gravity to three or four decimal places.

In this laboratory unit you will learn to prepare aqueous solutions using both methods described above. You will then measure the densities of the solutions you prepared and graph the density as a function of concentration to generate a standard curve from which the concentration of an unknown sample can be determined.

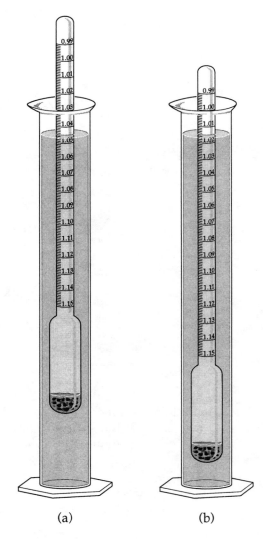

(a) (b)

Figure 15.7   (a) Hydrometer reading 1.045.
(b) Hydrometer reading 1.015.

PRE-LAB EXERCISES

1. Calculate the mass of sodium chloride you will need to make 1.00 L of a 1.000 M solution in part 1 of the procedure.

2. Calculate the volumes needed to prepare the diluted sodium chloride solutions you will make in part 2 of the procedure. Remember your handy formula:

$$C_{stock}V_{stock} = C_{dilute}V_{dilute}$$

An example calculation is provided below.

Example Calculation

To make 100.00 mL of a 0.50 M NaCl solution from the 1.00 M NaCl solution:

$$V_{dilute} = 100.0 \text{ mL}$$

$$C_{dilute} = 0.50 \text{ M}$$

$$C_{stock} = 1.00 \text{ M}$$

$$C_{stock}V_{stock} = C_{dilute}V_{dilute}$$

$$(1.00 \text{ M})(V_{stock}) = (0.50 \text{ M})(100.0 \text{ mL})$$

$$V_{stock} = \frac{(0.50 \text{ M})(100.0 \text{ mL})}{1.00 \text{ M}} = 50.0 \text{ mL}$$

So you would take 50.0 mL of the 1.00 M stock solution and dilute it to 100.0 mL.

PROCEDURE

1. Preparation of a 1 M NaCl Solution

Weigh out the mass of sodium chloride you calculated from the pre-lab exercise. Use a powder funnel to add it to a 1-L volumetric flask. Add enough distilled water to fill the flask about half full.

Swirl the flask contents gently until the solute has dissolved. Add water until the water level is at the meniscus. During the addition of the last few milliliters, you will want to use a pipet or dropper to add solvent, so that you won't accidentally overshoot the meniscus.

2. Serial Dilutions

Using the 1 M solution of NaCl you prepared above, prepare the solutions listed below.

- 100.00 mL of a 0.50 M NaCl solution from the 1.0 M NaCl solution
- 100.00 mL of a 0.25 M NaCl solution from the 0.50 M NaCl solution
- 100.00 mL of a 0.125 M NaCl solution from the 0.25 M NaCl solution

Have your instructor check your pre-lab calculations before proceeding with the dilutions.

3. Determination of the Densities of NaCl Solutions

Pour 50 mL of each NaCl solution into a 50-mL graduated cylinder. Use the hydrometer to determine the specific gravity of each NaCl solution. Record the value in your notebook. (Repeat this for each NaCl solution: 1.0 M, 0.50 M, 0.25 M, and 0.125 M.)

Determine the specific gravity of a 50-mL sample of water using the hydrometer. Record the value and the room temperature.

Your instructor will provide you with a reference value for the density of water. Use this value and the specific gravities you measured to calculate the densities of the NaCl solutions.

Obtain a sample of salt solution of unknown concentration from your instructor. Follow the procedure above to determine its specific gravity.

REPORT

1. Prepare a data table that lists the concentrations of the various solutions and their corresponding densities.

2. Either by hand or using a computer graphing program make a calibration curve: plot the density values (y-axis) as a function of concentration (x-axis). Draw the "best" straight line through your points. (If the points don't fall perfectly on a line, draw the line such that an equal number of points fall above and below the line. If you are using a computer program, it can do a "linear regression analysis" for you to draw the best line.)

3. On your graph, locate the density value for your unknown solution. From that point on the density axis (y-axis), draw a horizontal line to the calibration line. Then draw a perpendicular line from the calibration line to the x-axis to determine the concentration value. (See Figure 15.8).

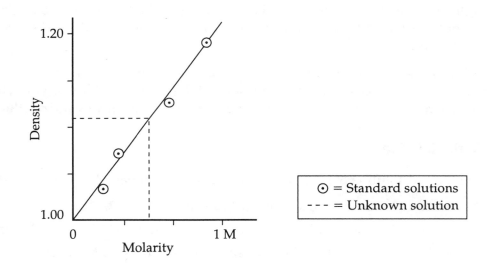

Figure 15.8   Sample graph showing the relationship between density and the molarity of aqueous NaCl solutions.

16

Kinetics: The Rate of a Reaction

OBJECTIVES

- Study the kinetics of the oxidation of iodide ion (I^-) by persulfate ion ($S_2O_8^{2-}$).

- Determine the reaction order of this oxidation with respect to iodide and persulfate, and use this information to postulate about the mechanism of this reaction.

- Observe how changes in temperature affect the reaction rate.

- Observe how addition of a catalyst affects the reaction rate.

EQUIPMENT/MATERIALS

Test tubes or small beakers, solutions of $Na_2S_2O_3$ in 0.4% starch, KI, KCl, K_2SO_4, $K_2S_2O_8$, and $CuSO_4$, large beakers, 10-mL graduated cylinder, ice, hot plate or Bunsen burner, stopwatch or clock/watch with a second hand.

INTRODUCTION

It is recommended that you review the section in your textbook that covers reaction rates and chemical kinetics.

Can you think of a chemical reaction that happens very fast? The oxidation (burning) of methane is an example. Once you apply a spark to a mixture of methane and oxygen, the reaction is virtually instantaneous. Think of a reaction that happens very slowly, such as the precipitation of calcium carbonate from water to make stalactites in a cave or the fossilization of organic matter. What determines whether a reaction occurs in the blink of an eye or so slowly as to be undetectable over a human lifetime?

Being able to control the path and rate of a chemical reaction can be advantageous. As an example, consider the decomposition of ammonium nitrate, a major component of many fertilizers. When heated, ammonium nitrate will decompose to form N_2O and water:

$$NH_4NO_3(s) \longrightarrow N_2O(g) + 2\,H_2O(g)$$

This reaction generates a fair amount of heat, and if that heat is not properly dispersed, it can generate higher temperatures at which the ammonium nitrate will explode:

$$2 \, NH_4NO_3(s) \longrightarrow 2 \, N_2(g) + O_2(g) + 4 \, H_2O(g)$$

This reaction should be avoided! Fertilizer manufacturers, shippers, and agricultural workers in particular will want to prevent the second of these reactions.

Studying how and how fast chemicals react can give chemists a lot of information about what is actually occurring at the molecular level in a chemical reaction. For example, studying the kinetics of biological processes helps biochemists design drugs that can enhance beneficial reactions or inhibit undesirable ones. Let's look at the reaction you will explore in this experiment, iodide ion being oxidized to iodine by persulfate ion:

$$2 \, I^-(aq) + S_2O_8{}^{2-}(aq) \longrightarrow I_2(aq) + 2 \, SO_4{}^{2-}(aq)$$

It is unlikely, though not impossible, that all three of the reactant ions will simply come together, react, and form product—a *termolecular* (three-molecule) process. A stepwise process with uni- or bimolecular reactions is more likely, but what are the actual steps that occur? By studying how the reaction rate is affected by changes in reagent concentration and temperature, we can gain insight into what is happening at the molecular level.

One possible sequence of steps would be for a single iodide ion to react with persulfate to form an *intermediate* species, which then goes on to react with a second iodide to form product, as described below.

Mechanism I

Step 1: $I^- + S_2O_8{}^{2-} \longrightarrow SO_4I^- + SO_4{}^{2-}$

Step 2: $SO_4I^- + I^- \longrightarrow I_2 + SO_4{}^{2-}$

Either step 1 or step 2 can be the slowest step. Remember, it is the slowest step that is *rate-determining*; that is, it determines the rate of the entire reaction, and the rate expression of the overall reaction reflects what is happening in the slowest step. For mechanism I above, two rate expressions are possible, depending on which step is rate-determining:

Rate expression if the first step is rate-determining: Rate $= k[I^-][S_2O_8{}^{2-}]$

Rate expression if the second step is rate-determining: Rate $= k[I^-][SO_4I^-]$

The second of these is difficult to verify directly because the concentration of the intermediate (SO_4I^-) is fleeting and hard to measure. However, through a series of approximations (the mathematical derivations of which we will spare you!), one can obtain a mathematical rate expression for the second step being rate-determining which doesn't contain the concentration of intermediate as a variable:

Approximation of rate expression if the second step is rate-determining:
Rate $= k[I^-]^2[S_2O_8{}^{2-}]$

This is also the rate expression for the situation in which all three ions react together in one single step (termolecular process).

A second possible mechanism, described below, involves the reaction of iodide ion with itself to make a different intermediate (I_2^{2-}), followed by reaction of the intermediate with persulfate ion.

Mechanism II

Step 1: $I^- + I^- \longrightarrow I_2^{2-}$

Step 2: $I_2^{2-} + S_2O_8^{2-} \longrightarrow I_2 + SO_4^{2-}$

In this case, the rate expression will be very different from mechanism I if the first step is rate-determining.

Rate expression if the first step is rate-determining: Rate $= k[I^-]^2$

If the second step is rate-determining, we are back in the same situation as above where the rate expression needs to be approximated and is indistinguishable from the termolecular process in which all three species come together and react in a single step:

Approximation of rate expression if the second step is rate-determining:
Rate $= k[I^-]^2[S_2O_8^{2-}]$

All of these possibilities are summarized in the table below.

Mechanism	Rate-Determining Step	Predicted Rate Law
I	1	Rate $= k[I^-][S_2O_8^{2-}]$
I	2	Rate $= k[I^-]^2[S_2O_8^{2-}]$
II	1	Rate $= k[I^-]^2$
II	2	Rate $= k[I^-]^2[S_2O_8^{2-}]$

Notice that both mechanisms have the same rate expression when the second step is rate-determining.

If we can figure out which rate law applies to this reaction, we can at least rule out some of the possibilities above. How, then, does one determine the rate law of a reaction? There are numerous approaches. In this experiment, you will vary the concentrations of the different reactants and see what effect those changes have on the reaction rate. Let's look at each of the three possible rate laws from the table above when we change concentration of reagents.

1. **Rate $= k[I^-][S_2O_8^{2-}]$**   In this case, if you double the concentration of I^-, the rate will double. Similarly, if you double the concentration of $S_2O_8^{2-}$, the rate will double, so the rate is *dependent on the concentrations of both reagents*.

2. **Rate $= k[I^-]^2[S_2O_8^{2-}]$**   In this case, if you double the concentration of I^-, the rate will increase by a factor of 4. As above, if you double the concentration of $S_2O_8^{2-}$, the rate will double. Thus, the rate *depends on the concentrations of both reagents* but is more sensitive to the I^- concentration.

3. **Rate $= k[I^-]^2$**   This rate law does not depend on the concentration of $S_2O_8^{2-}$ at all, so changing the $S_2O_8^{2-}$ concentration should have no effect on the rate. Conversely, doubling the concentration of I^- will increase the rate by a factor of 4. In this instance, the rate *depends only on the concentration of I^-*.

By adding starch to the solution, you will be able to detect the presence of I_2, which turns a dark blue-purple in the presence of starch. The faster the reaction is, the shorter the time required for the solution to turn purple. You will vary the concentrations of both I^- and $S_2O_8^{2-}$ to see what effect their concentrations have on the rate. In addition, you will explore the effect of temperature on reaction rate. Would you expect the reaction to occur faster or slower with increasing temperature? Why? You will also observe the effects of addition of a catalyst on the rate of this reaction.

PRE-LAB EXERCISE

Note: To vary the concentrations of I^- and $S_2O_8^{2-}$, you will use differing amounts of a stock solution. In order to keep the concentration of other reagents constant as well as maintaining the ionic strength of these solutions, you will be diluting with solutions that contain non-reactive ions, specifically KCl and K_2SO_4. For the purposes of this and other calculations, you can treat these solutions of nonreactive ions as if they were simply water (solvent).

1. Read through the procedure below. Prepare a data table in your notebook that lists the concentrations of the two reagents (I^- and $S_2O_8^{2-}$) in the final solution (assuming no reaction has yet taken place.) Leave a space to record the corresponding time it takes for the reaction mixture to turn purple. An example calculation is provided below.

2. Prepare a data table for the temperature study. List the concentrations for the reagents in the three temperature runs. (*Note*: You have already determined these concentrations through the calculation above.)

3. Consider the reaction $A + B \longrightarrow$ Products. The following two mechanisms are possible. The predicted rate expression is given for each mechanism.

Mechanism I	**Mechanism II**
$A \longrightarrow$ Intermediate (slow)	$A + B \longrightarrow$ Products (single step)
$B +$ Intermediate $\longrightarrow$ Products (fast)	$\text{Rate}_{II} = k_{II}[A][B]$
$\text{Rate}_I = k_I[A]$	

Given the following data, which mechanism would you expect to be operating?

$[A]_0$	$[B]_0$	**Time for reaction to go to 2% completion**[*]
0.10 M	0.10 M	57 seconds
0.20 M	0.10 M	30 seconds
0.10 M	0.20 M	59 seconds

[*]Remember, the faster the reaction, the less time it takes to go to completion.

Example Calculation for Concentration Study

In test tube B:
For I^-: $[I^-]_i = 0.20$ M $V_i = 3.0$ mL
$V_f = (3.0 \text{ mL KI}) + (1.0 \text{ mL KCl}) + (4.0 \text{ mL } K_2S_2O_8) = 8.0$ mL
$C_iV_i = C_fV_f$

$$[I^-]_f = \frac{[I^-]_i V_i}{V_f} = \frac{(0.20 \text{ M})(3.0 \text{ mL})}{8.0 \text{ mL}} = 0.075 \text{ M}$$

For $S_2O_8{}^{2-}$: $[S_2O_8{}^{2-}]_i = 0.10$ M $V_i = 4.0$ mL $V_f = 8.0$ mL
$$C_iV_i = C_fV_f$$
$$[S_2O_8{}^{2-}]_f = \frac{[S_2O_8{}^{2-}]_f V_i}{V_f} = \frac{(0.10\text{ M})(4.0\text{ mL})}{8.0\text{ mL}} = 0.050\text{ M}$$

PROCEDURE

1. Concentration Study

Obtain seven large test tubes and label them A–G. Place 2.0 mL of the solution of 0.0050 M $Na_2S_2O_3$ in 0.4% starch in each of the test tubes.

Place the appropriate amounts of solutions listed in the table below in the test tubes.

Test Tube	mL 0.20 M KI	mL 0.20 M KCl	mL 0.1 M K_2SO_4
A	4.0	0	0
B	3.0	1.0	0
C	2.0	2.0	0
D	1.0	3.0	0
E	4.0	0	1
F	4.0	0	2
G	4.0	0	3

Obtain ~35 mL of the 0.10 M solution of $K_2S_2O_8$. Use your graduated cylinder to measure 4.0 mL of $K_2S_2O_8$ solution. Add it to test tube A, stir the contents rapidly, and begin timing. When the solution turns blue-purple, record the time that has elapsed. (Your instructor may prefer that you perform the reaction in beakers to ensure better and quicker mixing.) When the reaction is finished, record the final temperature of the solution (for use in the temperature study below).

Repeat this process for test tubes B–G, adding the amounts of $K_2S_2O_8$ indicated in the table below. You do not need to record the final temperatures for B–G.

Test Tube	mL 0.10 M $K_2S_2O_8$
A	4.0
B	4.0
C	4.0
D	4.0
E	3.0
F	2.0
G	1.0

2. Temperature Study

You will use the rate data from the reaction of test tube A above for part of this study.

Obtain three large test tubes and label them H–J. Into each test tube, place 4.0 mL of the 0.0050 M $Na_2S_2O_3$ in 0.4% starch and 4.0 mL of the 0.20 M solution of KI.

Place H in an ice water bath[1] (0–5°C).

Place I in a warm water bath (~35–40°C).

Place J in a very warm water bath (~50–55°C).

Do *not* heat the solution of $K_2S_2O_8$, because the $S_2O_8^{2-}$ could decompose.

Allow about 5–10 minutes for the contents of the tubes to reach the same temperature as the bath, then record the actual temperatures of each of the baths.

Measure 4.0 mL of the 0.10 M solution of $K_2S_2O_8$ with a 10-mL graduated cylinder.

Remove H from the bath and quickly add the 4.0 mL of 0.10 M $K_2S_2O_8$ solution, stir rapidly, and begin timing. When the solution turns purple, record the time. Record the final temperature of the solution.

Repeat this process for test tubes I and J.

3. Catalytic Study

Repeat the procedure for test tube A above, except before adding the 4.0 mL of $K_2S_2O_8$ solution, add 1 drop of the copper sulfate solution provided (0.10 M). Stir rapidly, and time the reaction as before. Record your results.

REPORT

1. Prepare a table that lists the initial concentrations of I^- and $S_2O_8^{2-}$ and their corresponding reaction times for test tubes A–G. Use your results to select a reasonable rate expression. Which of the mechanistic possibilities outlined in the introductory section can be ruled out? Which one(s) is (are) most likely? Use your experimental results to support your reasoning.

2. Prepare a table that lists the reaction times for each of the four temperatures (A, H–J). What effect does increasing or decreasing the temperature have on the reaction rate?

3. Suppose you accidentally heated the $K_2S_2O_8$ solution during the temperature study, and it began to decompose. What effect would this have on your experimental results? Your conclusions about the relationship between rate and temperature?

4. Report on your results in part 3. Does the $CuSO_4$ catalyze the reaction? Briefly explain.

---

[1] To prepare a water bath, simply place water in a beaker (appropriately sized such that the test tube will neither tip the beaker over nor itself spill) and cool with ice or heat to the desired temperature.

17

Qualitative Aspects of Equilibrium

OBJECTIVES

- Develop a qualitative understanding of the effects of concentration on the position of an equilibrium.

- Develop a qualitative understanding of the effects of temperature on the position of an equilibrium.

EQUIPMENT/MATERIALS

Solutions of nitric acid (HNO_3), sodium hydroxide (NaOH), saturated sodium chloride (NaCl), concentrated hydrochloric acid (HCl), iron(III) nitrate ($Fe(NO_3)_3$), potassium thiocyanate (KSCN), cobalt dichloride hexahydrate ($CoCl_2 \cdot 6H_2O$), test tubes, beakers, hot plate or Bunsen burner, water, ice.

INTRODUCTION

When molecules react, the concentrations of reactants and products change continuously until the system reaches what is known as *chemical equilibrium*. At equilibrium, the reaction appears to come to a stop as the concentrations of reactants and products remain unchanged. However, a chemical equilibrium is *dynamic* in that both forward and backward reactions are still occurring. They occur at the same rate, such that no changes in the concentrations of reactants or products are evident. A general reaction in equilibrium can be written as follows:

$$aA + bB \rightleftarrows cC + dD$$

Le Chatelier's principle states that *if you disturb a reaction at equilibrium, the reaction returns to equilibrium by shifting in such a direction as to partially undo the disturbance.*[1] This means that if

---

[1] Russo, S. and M. Silver. *Introductory Chemistry*, 2d ed., San Francisco: Benjamin-Cummings, 2002.

$$aA + bB \rightleftarrows cC + dD$$

1. Increase the concentration of A.

2. Forward reaction rate increases:

$$aA + bB \rightleftarrows cC + dD$$

3. Concentrations of A and B decrease as the concentrations of C and D increase.

4. Rate of forward reaction slows, as rate of reverse reaction increases:

$$aA + bB \rightleftarrows cC + dD$$

5. Eventually a new equilibrium is established:

$$aA + bB \rightleftarrows cC + dD$$

Figure 17.1   Changing the concentration of a species in equilibrium.

you perturb an equilibrium by changing one of its key features (e.g., concentration or temperature), the equilibrium will adjust to minimize that disturbance. For example, increasing the concentration of A in the reaction above will cause the equilibrium to shift away from A. The forward reaction rate will increase and A and B will be consumed as the concentrations of C and D increase. As the concentrations of C and D increase, the rate of the reverse reaction begins to increase, until the rate of the reverse reaction becomes equal to that of the forward reaction and equilibrium is once again established with a new set of concentrations. (See Figure 17.1.)

The effect of changing the temperature is somewhat different. If the reaction above were exothermic, heat would be released as the reaction progressed. As shown in Figure 17.2, adding heat (i.e., increasing the temperature) would shift the reaction away from heat to the reactant side. Lowering the temperature (removing heat) would shift the reaction in favor of the production of heat to compensate, so the forward reaction would occur faster than the reverse, until a new equilibrium is established. As Figure 17.2 shows, the reverse is true for an endothermic reaction (heat as reactant).

The relationship among concentrations of the reactants and products can be expressed quantitatively using the *equilibrium constant*, K_{eq}. The mathematical expression for the equilibrium constant for the reaction above is

$$K_{eq} = \frac{[C]^c [D]^d}{[A]^a [B]^b}$$

where [A], [B], [C], and [D] represent the concentrations of A, B, C, and D, and the exponents (lowercase a, b, etc.) represent the stoichiometric coefficients from the balanced equation. If you increase the concentration of A (disrupting the equilibrium), the concentrations of C and D will increase as A and B decrease until equilibrium is reestablished and the value for the equilibrium constant is restored. Changing the temperature has a different effect. Changing the temperature *changes the value of the equilibrium constant*—it is only constant at a constant temperature. One can then observe the concentrations of the reactants and products changing until the new value for K_{eq} is satisfied and the system is once again at equilibrium.

For an *exothermic* reaction (heat as product)

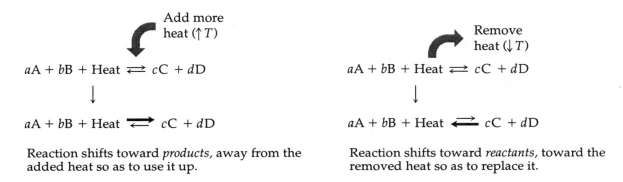

Add more
heat ($\uparrow T$)

aA + bB $\rightleftharpoons$ cC + dD + Heat

$\downarrow$

aA + bB $\xleftarrow{\rightharpoonup}$ cC + dD + Heat

Reaction shifts toward *reactants*, away from the added heat so as to use it up.

Add more
heat ($\uparrow T$)

aA + bB $\rightleftharpoons$ cC + dD + Heat

$\downarrow$

Remove
heat ($\downarrow T$)

aA + bB $\xrightarrow{\rightleftharpoons}$ cC + dD + Heat

Reaction shifts toward *products*, toward the removed heat so as to replace it.

For an *endothermic* reaction (heat as reactant)

Add more
heat ($\uparrow T$)

aA + bB + Heat $\rightleftharpoons$ cC + dD

$\downarrow$

aA + bB + Heat $\xrightarrow{\rightleftharpoons}$ cC + dD

Reaction shifts toward *products*, away from the added heat so as to use it up.

Remove
heat ($\downarrow T$)

aA + bB + Heat $\rightleftharpoons$ cC + dD

$\downarrow$

aA + bB + Heat $\xleftarrow{\rightharpoonup}$ cC + dD

Reaction shifts toward *reactants*, toward the removed heat so as to replace it.

Figure 17.2   Changing temperature in an equilibrium.

In this experiment you will study two different equilibrium reactions, one to observe the effects of concentration and the other to observe the effects of temperature. In both cases, the aim of the experiment is to obtain *qualitative* results so as to improve your understanding of the equilibrium process. Do not waste a lot of time being precise; the values given in the procedure are approximate and as long as you are close to what is recommended, you should get usable results.

Both of the equilibrium reactions you will explore involve the formation of a *complex ion*. A complex ion typically consists of a *metal ion* with *ligands* attached to it. Typical ligands are H_2O, NH_3, Cl^-, OH^-, CN^-, and SCN^-. The first reaction to be studied in this experiment involves an equilibrium among iron(III) ion (Fe^{3+}), thiocyanate ion (SCN^-), and the iron-thiocyanate complex ion, $(FeNCS)^{2+}$.

$$Fe^{3+}(aq) + SCN^-(aq) \rightleftharpoons (FeNCS)^{2+}(aq)$$

Fe^{3+} and SCN^- are both colorless in aqueous solution, but the iron-thiocyanate complex ion is a deep reddish orange color. How deeply the solution is colored gives you an idea of how much iron-thiocyanate complex ion is present and thus how far the equilibrium in the reaction above lies to the right.

The other reaction is an equilibrium between two metal complexes of cobalt:

$$\underset{\text{I}}{Co(H_2O)_6^{2+}} + 2\,Cl^- \rightleftharpoons \underset{\text{II}}{CoCl_2(H_2O)_2} + 4\,H_2O$$

When $CoCl_2 \cdot 6\,H_2O$ is dissolved under the right set of conditions, an equilibrium is observed between the hexa-aquo complex, I, and the bis-chloro-bis-aquo complex, II. One of these complexes is pink and the other is blue (and the difference is not subtle); furthermore, when a roughly equal mixture is present, the solution will be violet in color. You will use 95 % ethanol, which is 95% ethanol and 5% water, as your solvent. It is your job to determine which of the above complexes is pink and which is blue. You will also determine whether the equilibrium, as it is written above, is endothermic or exothermic.

PRE-LAB QUESTIONS

Consider the following equilibrium:

$$Ag^+(aq) + 2\,NH_3(aq) \rightleftarrows (Ag(NH_3)_2)^+(aq) + Heat\ (\Delta H = -15\ kJ)$$

How does each of the following situations affect the position of the equilibrium?

1. Dissolving $AgNO_3$ in the solution. Remember that metal nitrates are fully soluble in water and will dissociate upon dissolving: $M_x(NO_3)_y(s) \longrightarrow x\,M^{n+}(aq) + y\,NO_3^-(aq)$.
2. Bubbling $NH_3(g)$ into the solution.
3. Heating the solution.
4. Precipitating AgCl from the solution by adding NaCl:

$$Ag^+(aq) + Cl^-(aq) \longrightarrow AgCl(s)$$

PROCEDURE

1. Iron/Thiocyanate Equilibrium

Varying Fe^{3+} and thiocyanate concentrations.

Label five small test tubes, A, B, C, D, and E. Obtain ~5 mL of 0.5 M HNO_3 solution and ~2 mL each of the 0.020 M solution of iron(III) nitrate (ferric nitrate) in 0.5 M HNO_3 and 0.020 M solution of aqueous potassium thiocyanate. In each of the test tubes, place 0.5 mL of 0.5 M HNO_3.

Place the remaining solutions in the test tubes as indicated below.

- Test tube A: 1 drop potassium thiocyanate solution, 1 drop ferric nitrate solution
- Test tube B: 1 drop potassium thiocyanate solution, 9 drops ferric nitrate solution
- Test tube C: 5 drops potassium thiocyanate solution, 5 drops ferric nitrate solution
- Test tube D: 9 drops potassium thiocyanate solution, 1 drop ferric nitrate solution
- Test tube E: 9 drops potassium thiocyanate solution, 9 drops ferric nitrate solution

Compare the test tubes and record your observations.

Removing Fe³⁺

Label two small test tubes F and G. In test tube F, place 0.5 mL of 0.5 M HNO_3. In test tube G, place 0.5 mL of 1% NaOH (~0.25 M). Add 5 drops of potassium thiocyanate solution to both F and G. Add 5 drops of ferric nitrate solution to both F and G. Wait a moment and then compare the test tubes and record your observations.

2. Co²⁺ Equilibrium

Weigh out around 0.3 g of cobalt chloride hexahydrate, $CoCl_2 \cdot 6\,H_2O$, and place it in a clean, *dry* beaker. Dissolve it in 20 mL of 95% ethanol (95% ethanol contains about 5% water). Record your observations.

Place about 3 mL of the ethanolic cobalt chloride solution in each of six *dry* test tubes that you have labeled A–F. Test tube A will serve as your control for comparison.

Place test tube B in a hot water bath. (See Figure 17.3 and instructions below.) Place test tube C in an ice water bath. Wait a few minutes for them to equilibrate to the temperatures of the baths. Compare the tubes with the control (A) and record your observations.

Add 5 drops of a saturated aqueous sodium chloride solution to D. Observe the result and compare with A.

CAUTION: CONCENTRATED HYDROCHLORIC ACID IS CAUSTIC. AVOID CONTACT WITH SKIN AND CLOTHES. TREAT SPILLS WITH SATURATED SODIUM BICARBONATE SOLUTION AND COPIOUS AMOUNTS OF WATER.

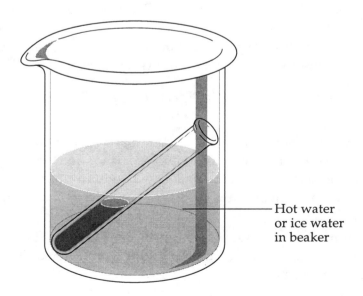

Hot water or ice water in beaker

Figure 17.3   Hot water or ice water bath.

Add 5 drops of a concentrated hydrochloric acid solution to E. Observe the result and compare with A.

Add 3 drops of distilled water to F. Compare with A and D.

Cool E in the ice water bath and compare with A and D.

Heat E in the hot water bath and compare with A and D.

Add an additional 3 drops of distilled water to E. Repeat the cooling/heating process, all the while comparing the solution to the solutions in A and D. Record all your observations.

Instructions for water baths: To prepare a *hot water bath*, choose a beaker of an appropriate size such that your test tube will stand up in it without tipping over. Fill the beaker about half full of water and heat the water with a Bunsen burner or hot plate. When the water is of the desired temperature, remove the heat and place the test tube inside. **Do *not* heat the bath while the test tube is in it, as ethanol is flammable.**

To prepare an *ice water bath*, choose the same size beaker as above and fill it half full of a slurry of ice in a small amount of water. (An ice water bath will cool something quicker than pure ice. Why?)

REPORT

1. Discuss your results from part 1 within the context of Le Chatelier's principle. Discuss the effect that adding and removing reactants or products (changing their concentrations) has on the position of the equilibrium based on your observations. The following reaction is *product-favored* and should be of use to you in interpreting your results:

$$Fe^{3+}(aq) + 3\ OH^-(aq) \longrightarrow Fe(OH)_3(s)$$

   Use this information to explain your observations of the position of the equilibrium in the presence of OH^-. (*Hint*: Think about what happens to $Fe^{3+}(aq)$ when there is $OH^-(aq)$ present.)

2. Discuss your results from part 2 within the context of Le Chatelier's principle. Look at the cobalt equilibrium reaction and decide which is pink (I or II) and which is blue.

$$\underset{I}{Co(H_2O)_6^{2+}} + 2\ Cl^- \rightleftarrows \underset{II}{CoCl_2(H_2O)_2} + 4\ H_2O$$

   Discuss the effect of changing the temperature based on your observations. Is the cobalt equilibrium exothermic or endothermic, as it is written? Explain your observations with respect to the addition of saturated aqueous sodium chloride solution, concentrated HCl, and the addition of water. (*Hint*: What is the major component of aqueous sodium chloride?)

3. Why is it important that your glassware be dry for part 2? Postulate as to what would have happened (i.e., how would your results have differed) if you had used water as your solvent in part 2?

18

What Color Is Red Cabbage?
A pH Exercise

OBJECTIVES

- Introduce some basic concepts relating to acid–base chemistry.
- Determine the relationship between an indicator's color and pH values.
- Determine the pH and acidity level of common household solutions.

EQUIPMENT/MATERIALS

Red cabbage leaves, 250-mL beaker, water, 10-mL graduated cylinder, 1 M KCl solution, standard pH solutions (pH 1–14), household substances.

INTRODUCTION

Acids and bases are among the most common chemicals you encounter. Cleaning supplies such as soaps, cleansers, drain openers, and glass cleaners are basic (low acidity) or alkaline. Baking powder, baking soda, and antacids are also bases. Car batteries contain sulfuric acid, and vinegar (acetic acid) is often used as a cleansing agent as well as a food item. Shampoos and cosmetics are usually pH-balanced to maintain a level of acidity that optimizes their performance and complements the pH of skin and hair. In this laboratory unit, you will examine the acid–base levels of various household items.

Acids are substances that can donate a hydrogen ion (H^+) and form H_3O^+ (hydronium ion) in water. Acids that release their H^+ easily and ionize completely in water are called *strong acids*. A solution of a strong acid will contain water molecules and ions only; no acid molecules will remain intact in solution (see Figure 18.1). It is 100% ionized.

$$HA_{strong} + H_2O \longrightarrow A^- + H_3O^+$$
$$\phantom{HA_{strong}}0\% 100\%$$

Many texts simply use H^+ instead of H_3O^+. The two terms are effectively interchangeable, but keep in mind that protons do not float around freely in water; they are associated with a

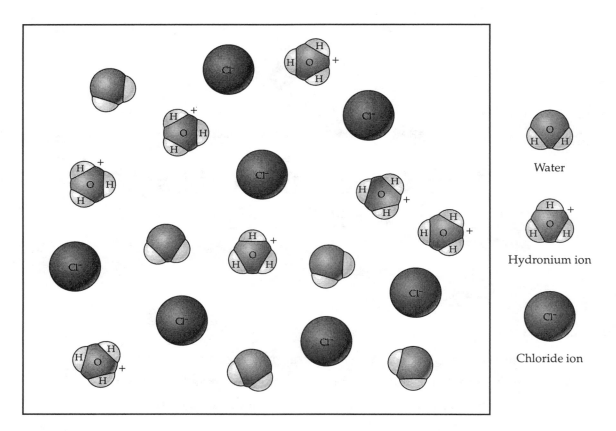

Figure 18.1   A solution of hydrochloric acid (HCl), a strong acid.

water molecule. Many strong acids are inorganic acids that are easily recognized by their chemical formulas. They usually start with H followed by nonmetallic simple or polyatomic anions (e.g., HCl, HBr, or HNO_3).

Acids that keep their H^+ most of the time and produce few ions in water are called *weak acids*. A solution of a weak acid will contain water molecules, undissociated acid molecules, and ions (see Figure 18.2). Formic acid (HCO_2H), for example, is only 4% dissociated in a 0.1 M solution and carbonic acid (H_2CO_3) is less than 1% dissociated in a 0.1 M solution. That means that in a formic acid solution, 4 out of every 100 formic acid molecules have released their hydrogen and are ionized, and in a carbonic acid solution, fewer than 1 out of every 100 carbonic acid molecules have released their hydrogen:

$$HCO_2H + H_2O \rightleftharpoons HCO_2^- + H_3O^+$$
$$96\% 4\%$$

In both cases, the solution contains mostly acid molecules and water molecules and only a very small number of ions.

In studying acids and bases, we encounter a semantic issue. In the vernacular, we might consider the terms *strong* and *concentrated* to mean the same thing; however, chemists distinguish between the terms *strong/weak* versus *dilute/concentrated*.

1. A *strong* acid is fully ionized in water.

2. A *concentrated* acid has a large amount of the acid molecule in solution.

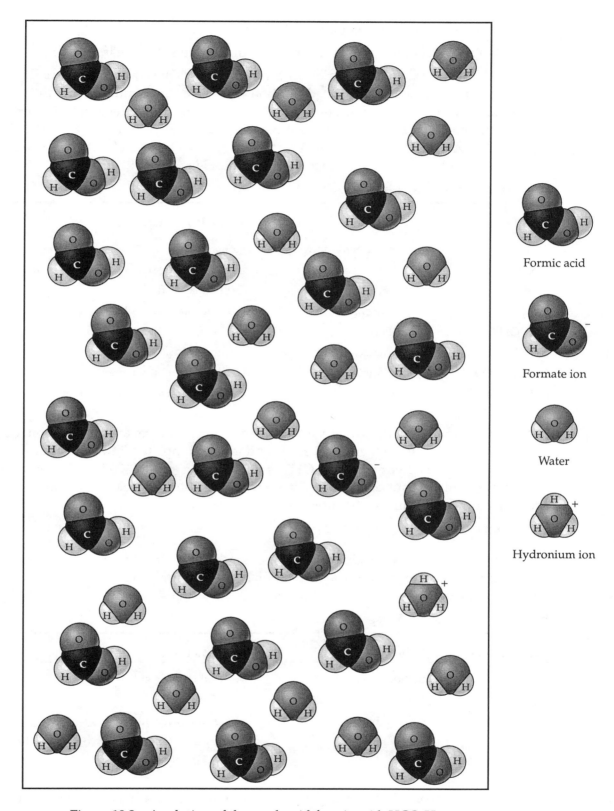

Formic acid

Formate ion

Water

Hydronium ion

Figure 18.2   A solution of the weak acid formic acid, HCO_2H.

3. A *weak* acid is only slightly ionized in water.

4. A *dilute* acid has a low concentration of the acid molecule in solution.

Acids and bases are found in a variety of concentrations. Thus, one may have a dilute solution of a strong acid or a concentrated solution of a weak acid. Similarly, one may have a concentrated strong acid or a dilute weak acid. Make sure you understand this distinction.

Organic acids are weak acids that contain a carboxyl group, $-CO_2H$, in the chemical formula. Carboxyl groups can donate protons to form $-CO_2^-$ and H^+. With the exception of stomach acid, all acids found in the body are weak acids.

Examples of Common Acids

Acid Formula	Name	Where It's Found
HCl	Hydrochloric acid	Stomach acid
HNO_3	Nitric acid	Explosives; once used to remove warts
H_2SO_4	Sulfuric acid	Car battery acid
H_3PO_4	Phosphoric acid	Used in cola soft drinks
HCOOH	Formic acid	Causes fire ant's sting
CH_3COOH	Acetic acid	Vinegar
$CH_3CH_2CH_2COOH$	Butyric acid	Causes rancid butter smell
$CH_3CHOHCOOH$	Lactic acid	Formed in muscles during exercise

Bases are substances that can accept a hydrogen ion, H^+, or form a hydroxide ion, OH^-, in water. *Strong bases* release their OH^- easily and ionize completely in water. They are usually identified by the hydroxide ion, OH^-, in the formula (e.g., NaOH, KOH). Some bases, such as ammonia, do not contain OH^- in their formula but release OH^- in solution by accepting a proton from a water molecule:

$$NH_3 + H_2O \rightleftarrows NH_4^+ + OH^-$$

Weak bases produce few ions in water. Bases found in the body are relatively weak and do not contain the hydroxide ion. The distinction between strong and concentrated or weak and dilute can be made in the case of bases as well as acids.

Examples of Common Bases

Base Formula	Name	Where It's Found
NaOH	Sodium hydroxide	Lye; used in drain cleaners
$Mg(OH)_2$	Magnesium hydroxide	Milk of magnesia
$NaHCO_3$	Sodium bicarbonate	Baking soda; antacid
NH_3	Ammonia	Cleaning agent; used to revive person who has fainted

The concentration of H^+ (in moles per liter) in a solution is described using the term *pH*. The pH of a solution is mathematically defined as follows:

$$pH = -\log[H^+]$$

In words, the pH of a solution is equal to the negative log of the concentration of H^+ (in moles per liter). The logarithmic relationship means that every time the pH increases by 1, the H^+ ion concentration increases by a factor of 10. The table below should help you to grasp this concept. Notice that a lower pH value corresponds to a higher H^+ concentration. A pH above 7 is basic, and a pH below 7 is acidic. As pH increases in value, the solution becomes less acidic and more basic. In a neutral solution, the pH is 7 and there are equal concentrations of H^+ and OH^- ions.

H^+ (moles per liter)	H^+ Concentration (in scientific notation)	pH	
1	1×10^0	0	↑
0.1	1×10^{-1}	1	A
0.01	1×10^{-2}	2	c
0.001	1×10^{-3}	3	i
0.0001	1×10^{-4}	4	d
0.00001	1×10^{-5}	5	i
0.000001	1×10^{-6}	6	c
0.0000001	1×10^{-7}	7 (neutral)	
0.00000001	1×10^{-8}	8	
0.000000001	1×10^{-9}	9	B
0.0000000001	1×10^{-10}	10	a
0.00000000001	1×10^{-11}	11	s
0.000000000001	1×10^{-12}	12	i
0.0000000000001	1×10^{-13}	13	c
0.00000000000001	1×10^{-14}	14	↓

Many naturally occurring molecules are influenced by the level of acidity around them. For example, litmus is a vegetable dye that changes color depending on acidity level. Molecules called anthocyanins are responsible for many of the red, blue, and purple colors of fruits, vegetables, and flowers. The color of these molecules is determined by the pH of their environment. The hydrangea flower has a red or pink color in acidic soil, but the same flower will be blue in alkaline or basic soil. Molecules that change color in response to changes in acidity level are called *indicators*. Indicators can be used to show when the pH of a solution has changed or to measure the pH of a solution. For example, phenolphthalein is colorless at a pH below 8 but turns pink in solutions whose pH is greater than 8.

$$\text{Ind-H} \quad \rightleftharpoons \quad H^+ + Ind^-$$

Indicator in acid form	Indicator in basic form
Phenolphthalein = colorless	Phenolphthalein = red

Some indicators change color only at one particular pH, whereas others, called universal indicators, exhibit a number of different colors over a broad pH range. Red cabbage leaves contain an anthocyanin compound that is a universal indicator; in this laboratory unit you will examine the color changes it undergoes as the pH around it changes.

PRE-LAB EXERCISE

1. Construct a data table for collecting indicator color data at pH values 1 through 14. This is your table of standard solutions. You will need to list pH values in one direction and observations (color) in the other.

2. Construct a second data table that links color data and pH values. Your instructor will provide you a list of different household substances, including tap water, which you will analyze.

PROCEDURE

1. Extraction of Red Cabbage Indicator

The pH-sensitive compound in red cabbage leaves is easily extracted from the leaves by boiling them in water. The compound dissolves in the hot water and the aqueous solution can be used as a pH indicator.

Obtain about 12 grams of red cabbage leaves and tear them into pieces about 1 cm². Place them in a 250-mL beaker and cover them with water until they are just barely floating.

Heat the cabbage–water mixture on a hot plate until the water boils. Boil for 5–7 minutes, then carefully remove from the heat and allow to cool.

Place a coarse coffee filter inside a large powder funnel that is in turn placed inside a 250-mL Erlenmeyer flask (see Figure 8.3). Filter your cabbage solution through the coffee filter. Discard the leaves and save the liquid.

You will need approximately 50 mL of indicator solution for this experiment. If your solution contains less than that, dilute it with distilled water until you have at least 50 mL. Note the color of the cabbage extract and record it.

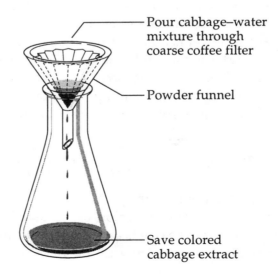

Pour cabbage–water mixture through coarse coffee filter

Powder funnel

Save colored cabbage extract

Figure 18.3   Experimental setup.

2. Determining the Indicator's Color at Different pH Values

Obtain a test-tube rack with 14 small test tubes. Label the test tubes 1 through 14 to correspond with the pH values you will use.

Use a 10-mL graduated cylinder to add 2 mL of the cabbage extract indicator to each of the 14 test tubes. Add 3 drops of 1 M KCl solution to each test tube. Mix the solutions gently.

Your instructor will direct you to the standard pH solutions. Add 2–3 mL of the standard solution labeled pH = 1 to test tube 1. Add 2–3 mL of the standard solution labeled pH = 2 to test tube 2, and so on.

Record the colors of the solutions in the test tubes in the data table that you prepared. Pay particular attention to the pH values at which significant color changes are evident (e.g., red → lavender, blue → green).

Save these test tubes to use as standards for comparison to the household substances you will test below.

3. Determining the pH Values of Household Substances

Obtain 10 additional test tubes and label them so that you know what household substance is in each.

Add 2 mL of the cabbage extract indicator and 3 drops of KCl to each of the 10 test tubes. Stir gently to mix.

Note any special instructions below before adding the household substances to the test tubes.

Record the color before and after mixing. After mixing, compare the solution to the standard solutions (from part 2 above) and, based on the color of the solution, assign a pH value that you think is the closest.

Repeat the procedure for each substance.

Organize your substances in order of *increasing* acidity based on your pH values.

Special Instructions

Note: If the substance is a liquid and is *not* listed below, add 2 mL of it to a test tube. If the substance is one of the following, add the specified volume:

Vinegar	1.0 mL
Lemon juice	1.0 mL
White wine	1.0 mL
Strong cleaning substance	0.5 mL

If the substance is viscous, like some shampoos, dip a stirring rod about ¼ inch into the substance, and use the rod to swirl the substance into a test tube that already contains the indicator and KCl. Be careful to thoroughly clean the stirring rod in between samples to avoid cross-contamination.

If the substance is a solid, grind it up in a mortar and pestle, if necessary, and add a small spatula tip of it to the test tube containing the indicator and KCl.

REPORT

Turn in your data tables as well as the answers to the questions below.

1. What pH corresponds most closely with the color of the original solution?

2. Can you make any statements about color trends for different pH values?

3. Which household substance is closest to neutral? The most acidic? The most basic?

4. Which substance has the highest H^+ concentration? The lowest H^+ concentration?

5. Categorize your household substances—for example, cosmetics, food, cleaning agents, and medicines. Do products that fall into the same category have similar pH values?

6. How does the pH of the food items tested compare to that of the cleaning agents tested?

7. How do the pH values of the tested medicine samples compare to one another and to the food items and cleaning agents tested?

19

Analysis of Commercial Antacids

OBJECTIVES

- Practice titration techniques.
- Observe the reaction between hydrochloric acid and a commercial antacid tablet.
- Determine the amount of acid neutralized by a commercial antacid tablet.
- Compare the effectiveness of a variety of commercial antacid tablets.

Review

- Pipetting techniques.
- Bar graph construction.
- Use of a buret.
- Indicator color and pH relationship.

EQUIPMENT/MATERIALS

Various commercial antacid tablets, stock solutions of approximately 0.100 M HCl and NaOH that have been standardized to known concentrations, phenolphthalein or bromothymol blue indicator, buret, buret stand, Erlenmeyer flasks, hot plate, stirring rod, mortar and pestle.

INTRODUCTION

In order to prevent the growth of bacteria, aid in the hydrolysis of certain foodstuffs, and denature most proteins, the cells of the human stomach walls secrete hydrochloric acid. The stomach has a pH of around 2 but is normally not harmed by the presence of the HCl, as it is protected by the mucosa, which continually replaces itself at a rate of approximately half a million cells per minute!

Occasionally the stomach responds to certain stimuli (stress, too much food, specific kinds of food, etc.) with an outpouring of acid, which lowers the pH to levels that cause discomfort (pH < 1). Some people experience even greater discomfort in the form of heartburn, which is a burning sensation caused when there is a backflow of stomach acid up into the esophagus. All antacids, regardless of claim or effectiveness, have one purpose: to neutralize the excess hydrogen ion in the stomach and/or esophagus and thus relieve acid indigestion. Commercial antacids all contain a base that reacts with excess stomach acid through a neutralization reaction. The effectiveness of an antacid depends on the amount of acid that it can neutralize. Various commercial products claim to give the "best relief" for acid indigestion.

Milk of magnesia, an aqueous suspension of $Mg(OH)_2$, is a simple antacid that neutralizes H^+ according to the following reaction:

$$Mg(OH)_2 + 2\,H^+ \longrightarrow Mg^{2+} + 2\,H_2O$$

The more common, "faster relief" antacids, which buffer[1] excess acid in the stomach, are those containing $CaCO_3$ or $NaHCO_3$. A H_2CO_3/HCO_3^- buffer system is established in the stomach when these antacids are used.

$$CO_3^{2-} + H^+ \longrightarrow HCO_3^-$$
$$HCO_3^- + H^+ \longrightarrow H_2O + CO_2(g)$$

The "seltzer" type of antacid also establishes a buffer system. These contain citric acid and sodium bicarbonate. These two substances do not react when they are both solids, but as soon as they come in contact with water, the following reactions occur:

$$H_3C_6H_3O_6 + HCO_3^- \longrightarrow H_2C_6H_3O_6^- + H_2O + CO_2(g)$$
$$H_2C_6H_3O_6^- + HCO_3^- \longrightarrow HC_6H_3O_6^{2-} + H_2O + CO_2(g)$$

The CO_2 gas that is given off accounts for the fizzing and foaming ("plop, plop, fizz, fizz!") exhibited by these products. The different conjugate bases of citric acid set up a very effective buffer.

Rolaids® and analogous products contain dihydroxyaluminum sodium carbonate, which is a combination antacid that neutralizes excess stomach acid as shown below.

$$NaAl(OH)_2CO_3 + 3\,H^+ \longrightarrow Na^+ + Al^{3+} + 2\,H_2O + HCO_3^-$$
$$HCO_3^- + H^+ \longrightarrow H_2O + CO_2(g)$$

In this experiment, you will determine the total effectiveness of different antacid brands using a strong acid–strong base *titration*. An acid–base titration is a method of determining the concentration of an acid or a base. The principle of a titration is fairly simple. You add one solution, whose concentration (of acid or base) is known, to another solution, whose concentration (of base or acid) is unknown. One mole of acid neutralizes one mole of base to form one mole of water, as shown below.

$$H^+ + OH^- \longrightarrow H_2O$$

[1] A buffer system is one that resists large changes in the pH of a solution when acid or base is added.

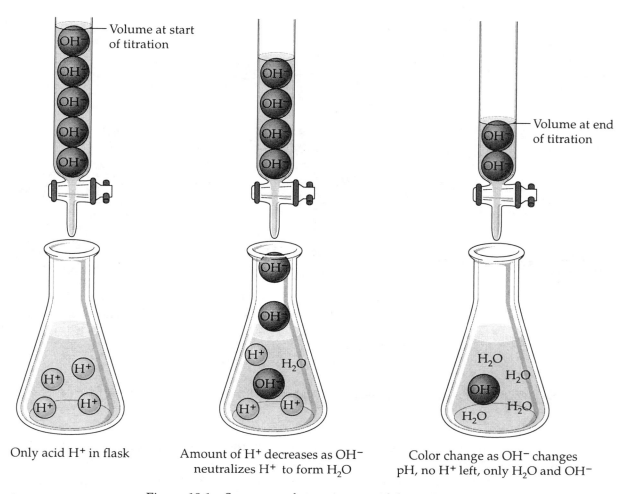

Only acid H⁺ in flask | Amount of H⁺ decreases as OH⁻ neutralizes H⁺ to form H₂O | Color change as OH⁻ changes pH, no H⁺ left, only H₂O and OH⁻

Volume at start of titration

Volume at end of titration

Figure 19.1   Sequence of steps in an acid–base titration.

As the acid and base neutralize each other, the pH of the solution gradually changes. When all moles of acid have been neutralized, any added basic solution will change the pH dramatically. This is called the *end point* of the titration. (See Figure 19.1.)

The end point of a titration can be monitored using a pH meter or a chemical *indicator*. An indicator is a substance—itself an acid or a base—that changes color depending on the pH of a solution. In this laboratory unit you will use either bromothymol blue or phenolphthalein as an indicator. Bromothymol blue is yellow in acidic solutions and turns blue in basic solutions. Phenolphthalein is colorless in acidic solution and turns pinkish red in basic solution. You can calculate the concentration of acid or base in the unknown solution by using the concentration and the volume of the known solution that was added.

$$\text{Concentration}_{acid} \times \text{Volume}_{acid} = \text{Concentration}_{base} \times \text{Volume}_{base}$$

To avoid the possibility of a buffer system[2] being established, excess HCl is added to the dissolved antacid, driving the equilibria in the neutralization reactions above far to the right.

---

[2]As mentioned above, buffers resist large changes in the pH of a solution. We must remove this buffering property to determine the antacid's total effectiveness. To do this, we swamp the system with excess strong acid and analyze for the amount of acid that was *not* neutralized by the antacid.

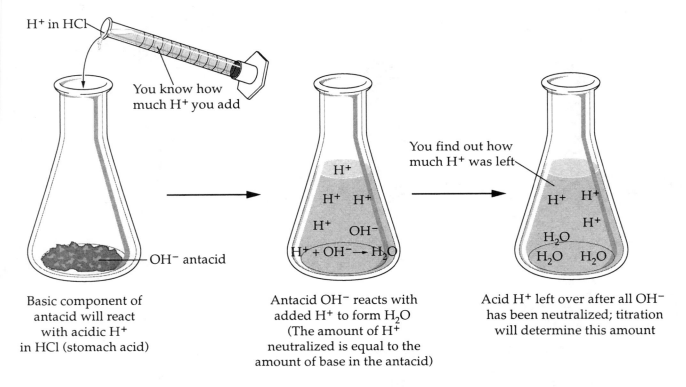

Figure 19.2　Sequence of steps involved in the neutralization of stomach acid (HCl) by the base in an antacid.

The solution is heated to drive off the CO_2 gas. Then the excess HCl is titrated with a standardized NaOH solution. An overview of this process is shown in Figure 19.2.

A neutralization reaction is one in which an acid and a base combine to form a salt and water. In species containing the bicarbonate ion (HCO_3^-) or the carbonate ion (CO_3^{2-}), carbon dioxide gas is also produced. The acidity (tendency to donate protons) and basicity (tendency to accept protons) of both substances are "neutralized" (rendered ineffective).

In this laboratory unit, you will simulate the reaction that occurs when antacids are taken to relieve excess stomach acid and compare the neutralizing capabilities of several commercial antacids. Mixing stomach acid in an Erlenmeyer flask with an antacid tablet simulates stomach activity. The base component of the antacid neutralizes the acid component of stomach fluid. Once the antacid's neutralizing capability is used up, there will be only acid (and some inert ingredients) left in the flask. Each antacid will neutralize a certain number of moles of acid, leaving some behind. The more effective an antacid, the fewer moles of acid will be left in the flask. You will figure out how many moles of acid are *left* by titrating the simulated stomach solution with a base of known concentration. Comparison of these figures will allow you to make a statement about which brand of antacid has the most neutralizing capacity.

PRE-LAB QUESTIONS

1. A 0.352-g sample of an antacid is dissolved in water. To it is added 40.00 mL of 0.141 M HCl solution. The solution is heated to drive off any CO_2 gas. It is then back-titrated to end point with 8.92 mL of a 0.203 M solution of NaOH.

a. How many moles of acid were in the original 40.00 mL of HCl?

b. How many moles of base were used in the titration of the HCl?

c. How many moles of acid were left in the flask after the antacid was added and the CO_2 was driven off (before the titration)?

d. How many moles of base were in the original sample of the antacid?

2. Prepare your notebook for this experiment. You will need to record the following information:

- Concentration of HCl used for all the trials
- Concentration of NaOH used for all the titrations
- Indicator used for all the titrations

For each trial, you will need to record the following information:

- Brand of antacid used and its active ingredient (from the product label)
- Exact mass of the antacid tablet used
- Volume of HCl added (usually 50.00 mL)
- Initial volume in buret before titration
- Final volume in buret after titration
- Any observations you make along the way

PROCEDURE

Record the exact concentration of HCl (approximately 0.1 M) and NaOH (approximately 0.1 M) that you will be using in your notebook.

1. Dissolving the Antacid and Adding Excess Acid

Do at least two trials for each antacid that you test. Your instructor will inform you of how many brands you are expected to analyze.

Obtain one tablet of an antacid. Crush the tablet with a mortar and pestle.

Weigh (to three decimals) a sample of a pulverized commercial-brand antacid tablet (about 0.3 g—but record the *exact* mass) and transfer the sample to a 250-mL Erlenmeyer flask.

Using a pipet, transfer 50.0 mL of HCl solution to the flask and swirl to dissolve. (Some of the inert ingredients in the tablet may not dissolve and/or the solution may appear cloudy.)

Heat the solution to a gentle boil and maintain the heat for 1–3 minutes to expel CO_2 gas.

Add 4–8 drops of bromothymol blue indicator[3] solution. Bromothymol blue is yellow in an acidic solution and blue in a basic solution. If the solution is blue, add an additional 25.0 mL of the HCl solution and boil again.

Cool the solution to room temperature by placing the flask under cold running water.

[3] *Note*: Phenolphthalein indicator may be used in place of bromothymol blue. Phenolphthalein is colorless in acid and pinkish red in base. If phenolphthalein is used, a pink color upon its addition indicates that more acid is needed to neutralize all the base present.

2. (Back)titrating the Sample

Standardized NaOH solution will be provided. Make sure that you have recorded the exact molarity that you use in your laboratory notebook.

Your instructor will demonstrate proper titration technique.

Clean a buret and fill it with standardized NaOH solution. Record the initial volume of NaOH solution. Titrate your sample solution from above to a blue (or pink) end point. The end point is reached when a blue color (without a greenish tint) persists for at least 30 seconds. Record the final volume of NaOH.

Refill the buret and repeat the experiment on a second sample of the same brand of antacid.

Perform the experiment, in duplicate, for as many different brands of antacid as your instructor assigns.

REPORT

Prepare the following tables to report your results. A sample calculation is provided below to guide you.

- *Table 1* lists each brand of antacid tested, its active ingredient(s), and the acid neutralization reaction.
- *Table 2* lists the mass of antacid in each trial, the moles of acid used, the moles of base required for titration, and the moles of base contained in the antacid.
- *Table 3* lists the brand of antacid and the moles of base per gram of tablet.

In addition, answer the following questions:

1. Which is the most effective antacid based on moles of base per gram of tablet? Which is the least effective?

2. If cost information is provided, which antacid is the most cost-effective (in terms of moles of base per dollar)?

3. If the CO_2 is not removed by boiling after the addition of HCl, how will this affect the amount of NaOH required to reach the titration end point? How will this affect (qualitatively) the determination of moles of base per gram of antacid tablet?

4. If the antacid selected for analysis is known to contain only milk of magnesia, how could this procedure be modified to expedite the analysis?

Example Calculation

Mass of antacid tablet used: 0.266 g

Concentration of HCl: 0.1011 M

Concentration of NaOH: 0.0997 M

Added 50.00 mL of HCl to antacid sample.

NaOH titration data: Final volume: 35.95 mL

$\underline{-\text{ Initial volume: } 29.02 \text{ mL}}$

Net volume: 6.93 mL

$$\text{mol}_{\text{NaOH}} \text{ (from the titration)} = (0.00693 \text{ L})(0.0997 \text{ mol/L}) = 6.91 \times 10^{-4} \text{ mol}_{\text{NaOH}}$$

Thus 6.91×10^{-4} mol $\text{HCl}_{\text{leftover after antacid neutralization}}$

$$\text{Initial mol (HCl)} = (0.05000 \text{ L})(0.1011 \text{ mol/L}) = 5.06 \times 10^{-3} \text{ mol HCl}_{\text{initial}}$$

$$\text{mol HCl}_{\text{neutralized by antacid}} = \text{mol HCl}_{\text{initial}} - \text{mol HCl}_{\text{leftover after antacid neutralization}}$$

$$\text{mol HCl}_{\text{neutralized by antacid}} = 5.06 \times 10^{-3} \text{ mol} - 0.691 \times 10^{-3} = 0.00436 \text{ mol}$$

$$\text{mol HCl}_{\text{neutralized by antacid}} = \text{mol}_{\text{antacid}} = 0.00436 \text{ mol}$$

$$\frac{0.00436 \text{ mol}}{0.266 \text{ g}} = 0.0164 \text{ mole of base per gram of tablet}$$

20

Analysis of the Chemical Components of Milk

OBJECTIVES

- Identify the main chemical components of milk.
- Gain experience in natural products separation.
- Apply some basic qualitative analytical tests.

EQUIPMENT/MATERIALS

Whole milk, acetic acid, coffee filters, ethyl acetate, Fehling's solution, Biuret's test solution ($CuSO_4$), xanthoproteic solution (nitric acid), lactase, sucrose (table sugar), glucose test solution, ammonium oxalate solution, ammonium molybdate solution, funnels, beakers, flasks, test tubes, thermometer, hot plate, watch glass.

INTRODUCTION

Got milk? Most of us know that milk and other dairy products provide a useful dietary source of calcium, which is important for bone and tooth growth and maintenance, as well as for numerous metabolic processes, including cell division, ATP synthesis, muscle contraction, nerve impulse transmission, and blood clotting. Milk is our first source of nourishment and has always played an important role in human nutrition. Indeed, the preparation of milk products such as cheese or yogurt predates recorded history. From a mammal's perspective, milk has to be an almost perfect food. For the first few weeks or months of life, milk must fill all of an infant mammal's nutritional needs. Each mammal's milk is uniquely suited to the metabolic and nutritional needs of that particular mammal.

In this experiment, you will isolate and identify some of the major components of cow's milk. The major components of milk are *water*, *protein*, *fat*, and the carbohydrate *lactose* (milk sugar). It also contains a variety of vitamins and minerals. The exact proportions of these components vary among types of milk (human, cow, goat, dolphin, etc.), among types of commercial varieties of cow's milk (skim, 2%, whole, etc.), and among brands. You will isolate some of the protein, some of the fat, and test for the presence of sugar.

Proteins

Protein molecules are composed primarily of amino acids linked together through amide bonds (also called peptide bonds). Each amino acid has at least one *amine* functional group ($-NH_2$) and one *carboxylic acid* functional group ($-COOH$). An amide bond results from a condensation reaction between these two functional groups, catalyzed in biochemical systems by enzymes. Enzymes themselves are proteins.

The surfaces of protein molecules display a large number of functional groups, represented in Figure 20.1 as R_1 and R_2. Many of these are amine or carboxylic acid functional groups. Amines act as bases. At normal physiological pH the amine groups are protonated and carry a positive charge, as shown in Figure 20.2. Conversely, the carboxylic acid functional groups become charged at high pH. They donate protons at normal physiological pH to become negatively charged (see Figure 20.3).

Many water-soluble proteins, including those found in milk, have an excess of either positive or negative surface charges due to their amino or carboxylate groups. As long as one charge or the other predominates, the protein will remain soluble. However, if the pH of the solution is changed dramatically (becoming either higher or lower, depending on the functional groups), so that the number of positive charges and the number of negative charges on the protein become equal, the protein will become less soluble because it will tend to aggregate (clump together) with other protein molecules. The pH at which this charge equality occurs is called the *isoelectric point*.

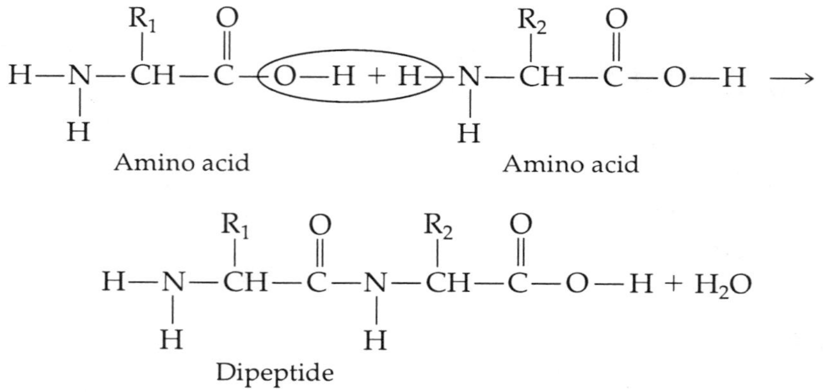

Figure 20.1   Two amino acids condensing to form a dipeptide. This repeats over and over until long chains of peptide residues are strung together to form a protein molecule.

$$R-NH_2 + H^+ \rightleftharpoons R-NH_3^+$$

High pH   Low pH

Figure 20.2   Amines accept protons to become cations as the solution pH falls.

$$R-CO_2H \rightleftharpoons R-CO_2^- + H^+$$

Low pH   High pH

Figure 20.3   Carboxylic acids donate protons to become anions as the pH rises.

A major protein constituent of milk, casein, can be precipitated from milk by acidification. By adding acetic acid until the final acid concentration is about 1%, the pH of milk (normally about pH 7) is lowered almost to the isoelectric point of casein, pH 4.6. Although casein usually goes by the single name, it is actually a mixture of four closely related proteins (α-, β-, γ-, and κ-casein), which you will not separate. Butterfat will precipitate with the protein and is separated with a bit of effort. Curdled solids and a sour taste are the hallmarks of spoiled milk. Both are caused by lactic acid, a by-product of bacterial growth in milk; the lactic acid also lowers the pH of milk, causing the casein to precipitate (curdle). Casein and fat are also the main constituents of the curd used in making cheese.

The other major proteins in milk are lactalbumin and lactoglobulin. These two proteins are soluble at both neutral and acidic pH. They do not precipitate in the curdling process and are often referred to as the "whey proteins." (Remember Little Miss Muffet and her curds and whey? The curds are the casein, discussed above; the whey is the liquid remaining after the curds have precipitated. The whey contains these whey proteins, as well as water, minerals, and carbohydrate.) Whey proteins can be precipitated by heating. Heat tends to unfold the compact structure of a soluble protein. The extended protein chains become entangled and form insoluble aggregates. The unfolding process is called denaturation and is usually irreversible. Heat denaturation of proteins is responsible for the "skin" on boiled milk and for the solidification of egg whites when they cook.

Fats

The fat found in milk is commonly called butterfat. It is suspended as globules in raw milk, but because the density of butterfat is less than that of water, it will rise and separate as cream. Homogenization is a process that reduces the average size of the fat globules so they will remain homogeneously mixed throughout the milk without separating out for long periods of time. Churning, on the other hand, is a method of causing the butterfat in cream to coagulate to a semisolid mass called butter, which is about 80% fat.

When the casein is precipitated by acidification of milk, most of the butterfat aggregates with it. The two can be separated based on their chemical differences. Fats are *esters* (another functional group) formed between long-chain carboxylic acids (fatty acids) and glycerol, a molecule having three carbon atoms, each possessing an –OH group. (See Figure 20.4.) The C_xH_y, C_mH_n, and other groups on the fatty acid may be the same or different.

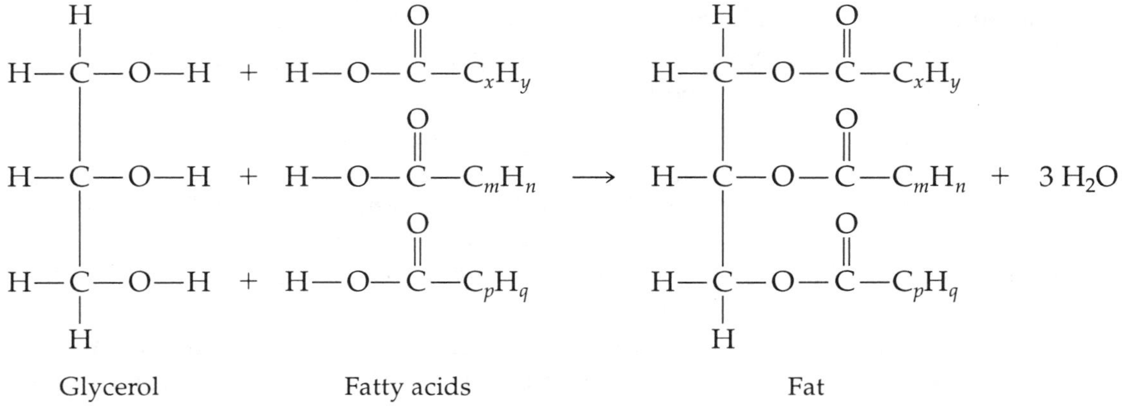

Glycerol   Fatty acids   Fat

Figure 20.4   The structures of a typical fat molecule and of glycerol.

They are typically long carbon chains (C6–20) with single or double bonds. Because milk contains animal fats, the chains are saturated, meaning they have no double bonds. Unlike proteins, fats do not have charged surface groups. Consequently, they are much more soluble in organic solvents and much less soluble in water. You will be able to separate the butterfat from the casein precipitate by extracting it with an organic solvent (ethyl acetate).

Carbohydrates

Carbohydrates are hydrated carbon chains, that is, carbon-containing molecules with –OH groups attached to the carbons. Sugars, starches, and cellulose are well-known hydrated carbon compounds. The carbohydrates found in milk are sugars, the major one being lactose. Lactose is a disaccharide, meaning it is composed of two covalently linked sugars, glucose and galactose, which are monosaccharides. Sucrose (table sugar) is another disaccharide, consisting of glucose covalently bonded to fructose. In order to digest lactose, the covalent bond between the two simple sugars must be cleaved. The enzyme lactase accomplishes this cleavage.

$$\text{Lactose (glucose–galactose)} \xrightarrow{\text{Lactase}} \text{Glucose} + \text{Galactose}$$

Lactase is produced by all mammalian infants. Adults of mammalian species, except for some humans, cease the ability to produce lactase and consequently are lactose-intolerant. The common symptoms of lactose intolerance in adults are gas, cramping, bloating, or diarrhea after consuming dairy products. This condition is different from a milk allergy, in which the proteins in milk trigger an allergic response. Lactose remains soluble in the whey even after acidification and heating. Its presence can be detected with a common test for sugars, the Fehling's test, which is not easily done at home. However, lactase can be purchased and when it acts on lactose, glucose is formed. Glucose tests (used by diabetics to monitor the urinary presence of glucose) are available commercially for home use.

Other Milk Constituents

Many vitamins, including A, much of the B complex, and C, are present in milk. Vitamin A is yellow in color and fat-soluble. It is primarily responsible for the color of butter and cheese. Most of the vitamin C is destroyed by heating during pasteurization. Vitamin D is not found naturally in milk, but it is commonly added. Among the more important mineral constituents found in milk are calcium and phosphate ions (Ca^{2+} and PO_4^{3-}). Combined with hydroxide ions, these two form hydroxyapatite, the major constituent of tooth enamel and the solid component of bone.

$$5\ Ca^{2+} + 3\ PO_4^{3-} + OH^- \rightleftharpoons Ca_5(PO_4)_3OH$$

The formation of hydroxyapatite from its ionic constituents is an equilibrium process, which normally favors the forward, or mineralization, reaction. The reverse reaction, demineralization, is favored by high acid content. Why?

Nursing mothers who lack sufficient amounts of these minerals in their diets are prone to loss of bone or tooth mass through demineralization. As calcium and phosphate ions are consumed in milk production, the blood level of these ions drops. Bone and tooth hydroxyapatite is normally stable relative to its constituent ions, but it will dissociate in response to low levels of these minerals in the blood (simple application of Le Chatelier's principle). Bacteria living on your teeth produce lactic acid, especially in response to high sugar levels. As acid levels rise, the rate of tooth demineralization increases. In whole milk, the calcium ions are mostly bound to protein.

PRE-LAB QUESTIONS

1. Prepare a flow chart in your notebook that shows all the steps of the procedure.
2. Most people who exhibit lactose intolerance do so because they lack the enzyme lactase. A very small percentage of people are lactose-intolerant because they are unable to metabolize the sugar galactose. Explain why taking a commercial preparation like Lactaid® will not help the second group of people.
3. Write chemical equations that show how an acidic environment will promote tooth decay.

PROCEDURAL OVERVIEW

- You will separate the casein and butterfat from the rest of the constituents by changing the pH of the milk (effectively curdling it).
- You will separate the casein from the butterfat by their different solubilities in an organic solvent.
- You will separate the lactalbumins and lactoglobulins from the rest of the constituents by heating.
- You will analyze the final filtrate for sugar (lactose) and minerals (calcium and phosphates). You will react the lactose with the enzyme lactase and analyze for glucose.
- You will confirm the presence of protein in the casein and whey proteins.

PROCEDURE

1. Isolation of Casein

Obtain a clean 250-mL Erlenmeyer flask. Measure out 100 mL of whole milk with a graduated cylinder and transfer the milk to the flask.

Place the flask in a water bath (a 600-mL beaker containing about 200 mL of water), and place the water bath on a hot plate. Heat the water to 40°C while carefully monitoring the temperature of the water with a thermometer. Once the water has reached 40°C, turn the heat off and give the milk sample about 5 minutes to warm to this temperature. Gently stirring or swirling the flask will ensure that the contents heat thoroughly.

Remove the flask from the water bath and add 10 to 15 mL of 10% acetic acid to the milk. Stir constantly for about 1 minute using a stirring rod. This will cause the casein and butterfat to precipitate out of solution.

Place three coffee filters, accordion-pleated together, in a large plastic or glass powder funnel. Place the funnel in a clean 250-mL Erlenmeyer flask and pour the mixture (curds and

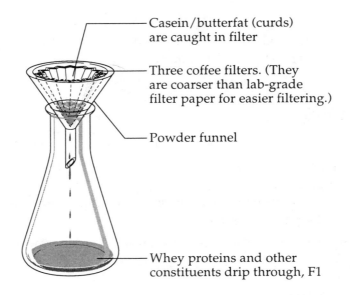

Casein/butterfat (curds)
are caught in filter

Three coffee filters. (They
are coarser than lab-grade
filter paper for easier filtering.)

Powder funnel

Whey proteins and other
constituents drip through, F1

Figure 20.5

whey) through the coffee filters very slowly, making sure the whey (liquid) is flowing into the flask. See Figure 20.5.

Use a spatula or rubber policeman to get as much of the precipitate into the filter as possible. While the substance is filtering, you may gently scrape the solid into a mound in the center of the filter. Be careful not to break the filter paper; this process takes time and you must use caution not to overflow the funnel. When the mixture is finished separating, label the filtrate (liquid) F1.

While the curds and whey are separating, set up a vacuum filtration apparatus according to Figure 20.6. Use a large (9-cm) Buchner funnel.

After the initial filtration is complete, gently lift out the coffee filters together containing the coagulated mixture and squeeze out the excess liquid, being careful not to break the filter. Spread out the coffee filters and use a scoopula to scrape off the mixture and transfer it to the Buchner funnel.

Break up the clumps and spread the curds across the filter. Turn on the vacuum and wash the curds with a stream of distilled water (10 to 20 mL should be plenty). Combine the filtrate from this washing with F1 and save it for the investigations of whey proteins (part 2). Allow the vacuum to run for about 5 minutes to dry the casein–butterfat mixture.

Transfer the dried curds to a dry coffee filter to check for moisture. If moisture is being absorbed by the coffee filter, scrape the residue onto a large watch glass. Heat the water bath to boiling and put the watch glass containing the precipitate on top to make a simple steam bath, as shown in Figure 20.7. As the heating progresses, water will separate from the precipitate. Remove as much of this extra liquid as possible, using a pipet, and discard it. Continue heating to get the residue as dry as possible. While the casein–butterfat mixture dries, proceed with the investigation of the whey filtrate, F1 (part 4 below). Be sure to check the drying process periodically and remove excess water as necessary.

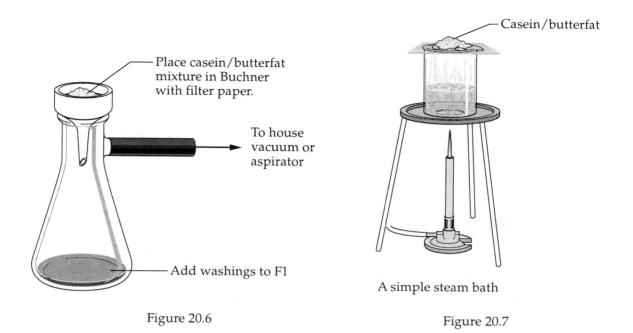

Place casein/butterfat mixture in Buchner with filter paper.

To house vacuum or aspirator

Add washings to F1

Figure 20.6

Casein/butterfat

A simple steam bath

Figure 20.7

2. Separation of Casein and Butterfat

Work in a hood or other well-ventilated area for this part of the exercise. Transfer your dry casein–butterfat residue from part 1 to a clean 100-mL beaker. Add 30 mL of ethyl acetate. Grind the residue with a glass rod to free as much of the trapped butterfat as possible. When no further dissolution of the residue is apparent, filter the suspension through a fluted filter secured in a funnel and placed over a large test tube containing a boiling chip. The filtrate collected in the test tube is F2 and will be used in part 3. Use a spatula or rubber policeman to make sure all the residue (casein) is transferred to the filter paper.

Place the filter paper containing the washed casein in a safe place in the hood. Spread it and the contents out and allow to air dry. Test the casein by the Biuret (part 6) and xanthoproteic (part 7) tests to verify that casein is a protein.

3. Butterfat Recovery

Work in the hood or a well-ventilated area. Place the test tube containing F2 in a hot water bath. (CAUTION!) Point the test tube to the side or back of the hood and boil until all the solvent has evaporated. Rigorous boiling is a characteristic of the solvent. Watch the color, and gently waft the odor to your nose to detect solvent. Boil off the solvent until the test tube is free of all solvent odor. Cool the tube and contents in an ice bath for at least 10 minutes. Note the consistency, odor, and appearance of the substance remaining in the test tube.

CAUTION: HOT PLATE ONLY, NO FLAME!

4. Whey Proteins

Place the beaker containing filtrate F1 into the water bath. Gently boil the filtrate to denature and precipitate lactalbumin and lactoglobulin. After the proteins coagulate, collect them by filtration. Save the filtrate, F3. Spread the collected whey proteins on a dry filter paper and allow them to air dry. Estimate the amount of whey proteins relative to the amount of casein by comparing their dry bulk. Carry out the Biuret (part 6) and xanthoproteic (part 7) tests on a portion of the whey precipitate to demonstrate that it is protein.

5. Tests on the Final Filtrate

The final filtrate, F3, should be tested for the presence of carbohydrate (lactose), calcium ions, and phosphate ions.

Fehling's Test for Carbohydrates

Add 5 mL of Fehling's reagent to a small test tube. Add 5 drops of F3 and boil the tube in water. A brick-red precipitate (Cu_2O) indicates the presence of an aldehyde, a functional group found in many sugars and most carbohydrates. The reaction involves the oxidation of the aldehyde group to a carboxylic group, as shown below.

$$RCHO + 2\,Cu(OH)_2 + NaOH \longrightarrow RCO_2^- + Na^+ + Cu_2O + 3\,H_2O$$

As a control (for comparison), run a parallel reaction using a 1% glucose solution as the carbohydrate.

Test for lactose

Obtain four small test tubes and label them 1–4.

Place ~0.5 mL of F3 in test tubes 1 and 2.

In test tubes 3 and 4, place a spatula tip of sucrose (table sugar) and dissolve it in ~0.5 mL of distilled water.

Grind up a lactase tablet using a mortar and pestle. Place a spatula tip of the resulting powder in test tubes 2 and 4. Gently swirl the test tubes or stir their contents with a stirring rod.

To summarize, your test tubes should contain the following substances:

1   0.5 mL F3
2   0.5 mL F3 + spatula tip of lactase
3   0.5 mL of water + spatula tip of sucrose
4   0.5 mL of water + spatula tip of sucrose + spatula tip of lactase

Follow the directions on the test strips for urinalysis provided (except you should test the solutions in test tubes 1–4, *not* your own urine!). Typically, you dip the test strip in the solution and wait 15 seconds to compare its color with the colors on the color chart provided. Use one test strip per solution and test each of the four solutions. Record your observations.

Test for Ca^{2+}

Add 1 mL of 0.1 M ammonium oxalate, $(NH_4)_2C_2O_4$, to 1 mL of F3. A white precipitate (calcium oxalate) indicates the presence of Ca^{2+}. Test an equal volume of untreated whole milk by this method. Compare the results.

Test for PO$_4$$^{3-}$

Add 2 mL of 6 M HNO_3 to 2 mL F3 in a medium test tube. Next add 2 mL of 0.5 M ammonium molybdate. Mix and then heat in a boiling water bath for 5 minutes. After heating, allow the test tube to stand at room temperature for at least 10 minutes. A positive test for phosphate will be indicated by the slow precipitation of ammonium phosphomolybdate, $(NH_4)_3PO_4 \cdot MoO_4$, a heteropolyanion compound. Describe the appearance of the precipitate.

6. The Biuret Test

Place one-quarter of a spatula of protein in a small test tube and add 15 drops of distilled water to suspend it. Add 5 drops of 10% NaOH and 2 drops of 0.5% $CuSO_4$, while swirling. Development of a violet or pinkish color indicates the presence of protein. Describe your result.

7. The Xanthoproteic Test

Place one-quarter of a spatula of protein in a small test tube and add 15 drops of distilled water to suspend it. Add 10 drops of concentrated HNO_3, while swirling.

> **CAUTION: CONCENTRATED NITRIC ACID IS STRONGLY CORROSIVE AND CAN CAUSE SEVERE BURNS.**

Heat the test tube carefully in a warm water bath. Development of a yellow color indicates the presence of protein.

REPORT

1. Write a brief summary of this experiment. Include a detailed flow chart showing all steps, precipitates/filtrates and their major composition, the test performed, and the results.

   a. Qualitatively compare the amounts of lactoglobulins and casein that you obtained. Describe the results of your protein analysis of the casein and the whey proteins.

   b. Describe the results of your butterfat recovery.

   c. Describe the results of your carbohydrate analysis. Include a table that reports the results of the glucose analysis of the different solutions in the test tubes in part 5b. Discuss these results, given what you know about lactose and lactase. What was present in each test tube? What evidence do you have?

   d. Describe the results of your mineral analysis (Ca^{2+} and PO$_4$$^{3-}$).

2. Answer the following questions:

   a. If the isoelectric point of casein is pH 4.6, which type of ionic group on the protein's surface predominates at pH 7 in milk, $-NH_3^+$ or $-CO_2^-$?

   b. Fluoridation of water helps prevent tooth decay because the F^- ion replaces the OH^- ion in hydroxyapatite. Briefly explain how this works. Would you expect fluoridation to improve bone strength as well? Why or why not?

21

An Introduction to Beer's Law

OBJECTIVES

- Develop an understanding of the relationship between the absorption of light and molecular concentration.
- Develop an appreciation for the necessity and use of a calibration curve.
- Examine the relationship between wavelength of light and color.
- Learn how to use the Spectronic-20.

EQUIPMENT/MATERIALS

Spec-20, cuvettes for Spec-20, dye solutions, copper wire, 10 M nitric acid solution, volumetric flask, beakers, flasks, pipets.

INTRODUCTION

Light can be thought of as energy traveling in waves through space. These waves can be described in terms of wavelength (the distance in space before the light wave repeats itself) or frequency (number of complete waves passing a particular point in space per unit time). Wavelength is given the symbol λ (Greek lambda), and is often reported in units of meters or nanometers (see Figure 21.1). Humans visually perceive a very narrow range of wavelengths, called visible light, between 750 and 400 nanometers (nm = 10^{-9} m). This visible spectrum encompasses all of the colors that we perceive, and, as can be seen in Table 21.1, each color has a particular wavelength region associated with it.

When light falls on any object, three things can happen:

1. Some may bounce off (be reflected).
2. Some may pass through (be transmitted).
3. Some may be absorbed.

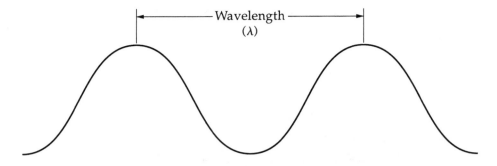

Figure 21.1   Wavelength of an electromagnetic wave.

Normally we view things using white light (light with a uniform distribution of wavelengths). If the object to be viewed absorbs all the light, it appears black. If the object reflects all the light, the object is said to be white. Most materials are not at either of these extremes but have a complex response to light resulting in some perceived color. A substance will appear a particular color because it absorbs visible light of every wavelength *except* that color. If an object appears to be green in white light, this means that all the colors except green are absorbed and only the green light is reflected (opaque green object) or transmitted (transparent green object). A colorless object transmits all wavelengths and is visible only because of differences in a property called refractive index.

The perceived color of an object depends on the properties of the light source and the properties of the material. Atoms and molecules can absorb light of different wavelengths, and the absorption of visible light is generally associated with changes in the energy of the electrons within atoms or molecules. The quantitative relationship between wavelength of light and absorption of that light by a substance is called the *absorption spectrum*. Figure 21.2 shows the absorption spectrum of both versions of the plant pigment chlorophyll, a and b. Both absorb strongly between 400 and 500 nm (blue/violet region) and between 600 and 700 nm (orange/red). Neither absorbs in the green region (~500–600 nm), which is why chlorophyll, and all the plants that contain it, *appear* green: the green light is not being absorbed but, rather, transmitted.

Table 21.1   Wavelength and Color

Color of Light Absorbed	Wavelength of Light Absorbed (nm)	Color of Light Transmitted
Violet	400–440	Yellow-green
Blue	440–480	Yellow
Green-blue	480–490	Red-orange
Blue-green	490–500	Orange
Green	500–560	Red
Yellow-green	560–580	Violet
Yellow	580–600	Blue
Orange	600–610	Blue-green
Red	610–750	Green

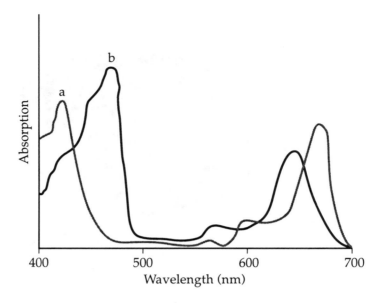

Figure 21.2 Absorption spectrum for chlorophyll a and b.

The absorption of light by a substance in a solution can be described mathematically by the *Beer–Lambert law* (often just called Beer's law):

$$A = \epsilon bc$$

In this equation,

> A = absorption by the colored species at a given wavelength of light; it has no units.

Greek epsilon, ϵ = molar absorptivity constant (also called the extinction coefficient), which is unique to each molecule and varying with wavelength; it has units of $L/M^{-1}cm^{-1}$.

> b = the path length or distance through the sample that the light has to travel and has units of cm.

> c = the concentration of the solution in moles per liter (M).

For absorption measurements of a given substance on a given instrument at a given wavelength, ϵ and b are constants. Thus, there is a proportional relationship between absorption of light and concentration. Mathematically this can be expressed as $A \propto c$ or $A = mc$, where m is a constant. That absorption and concentration should be directly proportional makes intuitive sense in that a concentrated solution of a dye will appear more deeply colored than a dilute solution.

Spectroscopy is the study of the interaction of electromagnetic radiation ("light") with various materials. We will use a simple absorption spectrophotometer as diagrammed in Figure 21.3 to study the relationship between (a) wavelength and color and (b) absorption and concentration. This instrument, the Spectronic-20, is nicknamed the "Spec-20."

How the Spec-20 Works

Inside the spectrophotometer a special light bulb produces white light, which is made into a focused beam with a set of slits. A diffraction grating (analogous to a prism) spreads the light

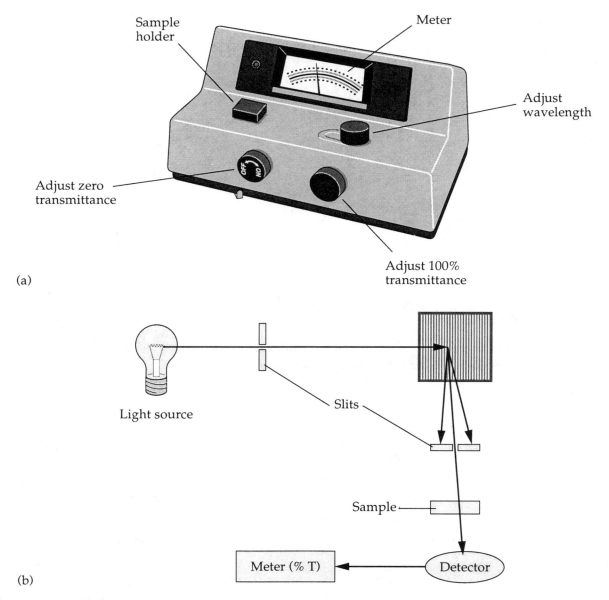

Figure 21.3   The Spectronic-20. (a) Actual instrument, (b) schematic diagram.

in space according to wavelength. The wavelength desired is selected by moving the grating, which you do by turning the wavelength selection knob so that a particular wavelength can come through the second slit. The light passes through the sample and on to the detector, where the amount of light that gets through is measured on a meter or a digital display.

The detector part of the Spec-20 compares the amount of light your sample transmits to the amount of light transmitted by a *blank*. The blank (which in our case is just pure water) is inserted first; the instrument is "zeroed" (i.e., you tell the detector that the blank is *not* absorbing any light). Then when you put the sample of colored solution into the Spec-20, the detector tells you how much the sample absorbs compared to the blank.

The scale or readout on the Spec-20 provides two types of values:

- % T (percent transmittance): the percent of light that is transmitted through the sample at a given wavelength compared to the reference.

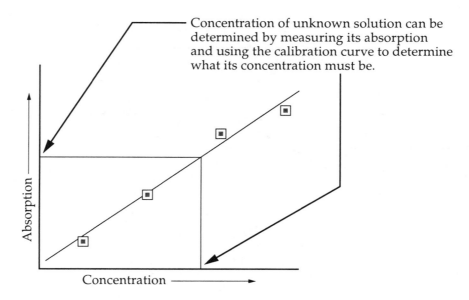

Concentration of unknown solution can be determined by measuring its absorption and using the calibration curve to determine what its concentration must be.

Figure 21.4   Beer's law plot.

- A (absorption): amount of light absorbed by the sample at a given wavelength compared to the reference.

Absorption and percent transmittance are related by the following equation:

$$A = \log \frac{1}{\% \, T}$$

If the absorption is measured for a series of solutions with known concentrations of absorbing molecules, a Beer's law plot, also called a *calibration curve*, can be generated. A calibration curve is illustrated in Figure 21.4; it is a plot of absorption as a function of concentration at a given wavelength. The more concentrated the solution is, the more light it will absorb, and this relationship is linear. Once the calibration curve is drawn, the concentration of an *unknown* sample may be determined by measuring the absorption of the sample at the same wavelength. You then use the absorption value to derive a corresponding concentration value from the curve. (See Figure 21.4.)

In this experiment, you will familiarize yourself with the Spec-20 spectrometer and use it to make two kinds of measurements. You will first measure the absorption of three dye solutions at different wavelengths of light and then graph absorption as a function of wavelength for these dyes. Second, you will measure the absorption of some known solutions of copper(II) and generate a calibration curve. You will use this calibration curve to determine the concentration of an unknown copper solution.

PRE-LAB EXERCISES

1. Write the balanced reaction that occurs when copper dissolves in nitric acid. (Your instructor may need to help you with this.)

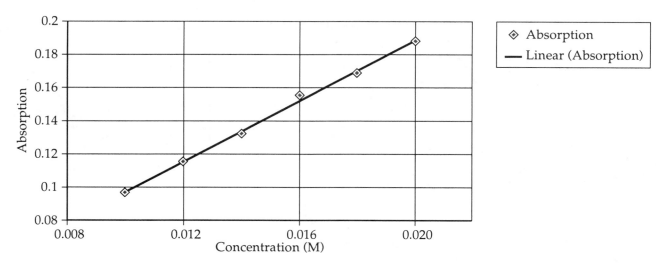

Calibration curve

Figure 21.5   Calibration curve.

2. If you dissolve 0.135 g of copper in nitric acid and dilute it to 100.0 mL with water, calculate the concentration of the resulting solution.

3. The calibration curve shown in Figure 21.5 relates concentration to absorption for a red dye. Use this calibration curve to determine the concentration of an unknown solution, if the absorption for the unknown is 0.15.

4. Prepare your notebook for this experiment. You will need tables to record the absorption of three dyes and 20-nm wavelength increments between 400 and 750 nm. You will need to record the masses of four copper samples and the absorptions of the corresponding solutions. You will also need to record the absorption of a copper solution of unknown concentration.

PROCEDURE

1. Dye Solutions

The dye solutions will already be prepared for you. Obtain about 20 mL each of the green, yellow, and blue dye solutions. Record the concentrations of the solutions (written on the bottles) in your notebook.

Obtain and thoroughly clean a set of cuvettes (fancy test tubes) for the Spec-20. Use soap solution to clean the cells but *do not use a brush* (to avoid scratching the cells). After the cells have been cleaned inside and out with soap solution, rinse them several times with tap water and then at least twice with distilled water.

Your instructor will provide detailed instructions for the operation of the Spec-20. Prepare a blank by filling a cuvette with distilled water and wiping the outside of it dry.

Set the wavelength dial on the Spec-20 to 400 nm. Adjust the instrument to read as follows:

- 0% T *with no cell present*
- 0 absorption when the distilled water-filled cell is in the sample holder

Rinse one of the cells several times with one of your dye solutions. Fill the cell with a fresh portion of the same solution and carefully wipe the outside of the cell dry with a lens tissue or Kimwipe®. After it has been dried, the lower half of the cell should not be handled—*fingerprints will cause incorrect readings.* Place the other dye solutions in cuvettes using a separate one for each and following the same rinsing, filling, and drying procedure used for the first solution.

Remove the distilled-water blank from the Spec-20 and insert the cell containing the dye solution. Read and record the absorption. Remove the sample cell. Rezero the spectrometer with the blank. Replace the first dye solution with the second and record the absorption. Rezero the spectrometer with the blank and measure the absorption of the third dye solution.

Measure and record the absorption of each of the dyes at 20-nm increments from 400 nm to 750 nm (i.e., at 400, 420, 440, etc.). *Remember that you need to rezero/restandardize the Spec-20 with your blank each time you change the wavelength.* Therefore, it would be most efficient to zero the instrument, measure the absorption of each of the three dyes at that wavelength, and then move on to the next wavelength, remembering to rezero. Just be sure to record your absorption measurements carefully so you do not get the dyes mixed up!

2. Copper Solutions

> **CAUTION: 10 M HNO_3 IS *EXTREMELY* CAUSTIC AND DANGEROUS. IT WILL BURN HOLES IN YOUR SKIN AND YOUR CLOTHES. AVOID BREATHING THE REDDISH BROWN NO_2 GAS THAT FUMES OFF THE ACID. IF YOU SPILL THE CONCENTRATED HNO_3, APPLY SATURATED BICARBONATE IMMEDIATELY.**

Use a balance to accurately weigh a copper wire sample in each of four clean, dry 150-mL beakers. Label the beakers A, B, C, and D. Record the mass of each wire used.

In the large fume hood, add 10 mL of 10 M HNO_3 to each sample **(CAUTION!)**. To facilitate the dissolution of the copper wire, it may be necessary to warm the beaker on a hot plate for 4–5 minutes. Add water as necessary to maintain a solution. **Do *not* boil to dryness.**

When the copper wire is *completely* dissolved, add 30 mL of distilled water and transfer the beaker contents quantitatively to a 100-mL volumetric flask. This is accomplished by rinsing the beaker thoroughly with several 10-mL portions of distilled water and adding each rinsing to the volumetric flask without loss of material.

If necessary, add distilled water to the flask until the liquid level is three-quarters of the way to the neck of the flask. Then thoroughly mix the solution by gentle swirling. After air bubbles are no longer visible in the solution, carefully add distilled water until the bottom of the meniscus is at the calibration mark on the flask. (You should add the water drop by drop with a dropper or wash bottle when the meniscus is about a centimeter away from the etched line.)

Stopper the flask, and mix the solution thoroughly by slowly inverting the flask at least ten times, each time allowing the air pocket to reach the bottom of the flask. Transfer the contents to a clean, dry container beaker or Erlenmeyer flask, and label it. Rinse out your volumetric flask thoroughly with distilled water. Repeat this procedure for each of the four samples. Calculate and record the molarity of each solution.

In some instances, it may be necessary to prepare a fifth sample, which is extremely dilute, in order to generate an accurate calibration curve to use with the unknown solution. Wait until you have tried to measure the absorption of your unknown before you decide to do this (your instructor can guide you). If you *do* need to prepare a fifth sample, withdraw 10 mL of the most concentrated solution and transfer it to your volumetric flask. Dilute the sample up to the 100-mL mark and then transfer the contents to a clean, dry beaker or flask and label it. This process is called a serial dilution.

Thoroughly clean your set of cuvettes. From the four copper solutions just prepared, select the most dilute. Rinse one of the cells several times with this solution of $Cu^{2+}(aq)$. Fill the cell with a fresh portion of the same solution and carefully wipe the outside of the cell dry with a lens tissue or Kimwipe®. After it has been dried, the lower half of the cell should not be handled—*fingerprints will cause incorrect readings*. Place the remaining $Cu^{2+}(aq)$ solutions in cells, using a separate cell for each and following the same rinsing, filling, and drying procedure used for the most dilute solution.

Prepare a blank by filling another cell with distilled water and wiping the outside of the cell dry. Set the wavelength dial on the Spec-20 to 640 nm. Adjust the instrument to read as follows:

- 0% T (upper scale) *with no cell present*

- 0 absorption (lower scale) when the distilled water-filled cell is in the sample holder

Remove the distilled-water blank and insert the cell containing the most dilute $Cu^{2+}(aq)$ solution. Read and record the absorption. Remove the sample cell. Rezero the spectrometer with the blank. Use this procedure to determine the absorption for each of the four (or five) solutions at least three times, being careful to zero the spectrometer between measurements.

Obtain a sample of an unknown $Cu^{2+}(aq)$ solution from your instructor. Record the number of the unknown in your notebook. Measure and record the absorption of the unknown at 640 nm, three times, remembering to zero the instrument with the blank between readings.

If your absorption readings for the unknown do not fall within the range of values you obtained for the calibration curve, you will need to prepare a more dilute standard solution (see above for instructions). Measure the absorption for the dilute standard solution, and then go back and measure the absorption (three times) for your unknown again.

REPORT

1. Prepare a graph of absorption as a function of wavelength for each of your three dyes. Your instructor may direct you to use a computer graphing program. Look at your three graphs (overlays might help) and postulate about the composition of the green dye. Discuss the difference between the appearance of the dye and the wavelength of light it absorbs.

2. Prepare a table of the concentration and corresponding absorption values of your known copper solutions. Include a sample calculation for the determination of the concentration.

3. Calculate an average value for your three absorption measurements for each of the copper solutions. Prepare a Beer's law calibration curve for $Cu^{2+}(aq)$.

4. Use your $Cu^{2+}(aq)$ calibration curve (Beer's law plot) to determine the concentration of your unknown copper solution. Briefly explain how you arrived at your answer for concentration. Calculate ϵ for $Cu^{2+}(aq)$ at 640 nm (assuming b = 1.00 cm).

22

Spectronic-20 Analysis of Aspirin Content in Commercial Products

OBJECTIVES

- Further explore Beer's law.
- Determine the concentration of aspirin in a commercial tablet using the Spec-20 and a Beer's law calibration graph.

Review

- Dilution preparations and calculations.
- Construction and use of a calibration curve.
- Use and operation of a Spec-20.

EQUIPMENT/MATERIALS

Standard aspirin solutions, 1.0-mL and 10.0-mL pipets, 10-mL Mohr pipets, burets, 16-cm test tubes, cuvettes, test-tube rack, Spec-20, commercial aspirin solutions, 0.10 M nitric acid solution, 0.10 M iron(III) solution.

INTRODUCTION

The previous laboratory unit describes how color results when visible light interacts with matter that absorbs some of the light and transmits the rest. The color of a substance depends on the wavelength of the light that is absorbed; the remaining wavelengths are transmitted, and this is the color we see. The amount of light absorbed depends on the number of light-absorbing species present in the substance. As we discussed in the previous experiment, Beer's law states that the more light-absorbing species there are in a solution, the more light the solution absorbs and the less it transmits (see Figure 22.1). The mathematical expression of Beer's law is

$$A = \epsilon bc$$

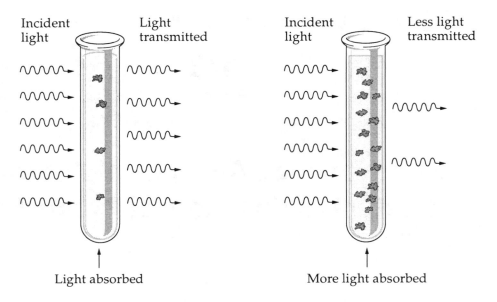

Figure 22.1   The amount of light absorbed and emitted depends on
the number of light-absorbing species.

where

A = absorption at a given wavelength of light,

Greek epsilon, ϵ = the molar absorptivity,

b = the path length, and

c = the concentration of the solution in moles per liter.

Remember that when you are performing spectroscopy with a constant path length at a constant wavelength, the absorption, A, is proportional to the concentration of the absorbing species (A = mc). By first constructing a calibration curve, you can determine the concentration of an unknown species by measuring its absorption and correlating it to the concentration. (See previous laboratory unit for a detailed explanation.)

In this laboratory unit, you will measure the absorption for a series of solutions that contain known concentrations of a colored species. You will plot a graph showing the relationship between the concentrations of your samples and the amount of light they absorb. From this graph, you can obtain the concentration of an unknown aspirin sample by measuring its absorption. The concentration unit used in this experiment is mg/mL instead of molarity.

Salicylates are a family of drugs that relieve pain and inflammation. (See Figure 22.2.) The best-known member of this family is acetylsalicylic acid, aspirin. Though normally colorless, when reacted with iron salts, salicylates take on a reddish violet color. This reaction can be used along with spectroscopic analysis to confirm suspected cases of salicylate poisoning by determining the concentration of salicylate in a blood sample. Commercial aspirin tablets may not be pure aspirin. They often contain inert filler ingredients to help maintain the integrity of the tablet (prevent crumbling) or other substances that either also provide pain relief or mitigate stomach irritation experienced by many consumers. In this lab, you will test and compare two commercial aspirin products for their aspirin content.

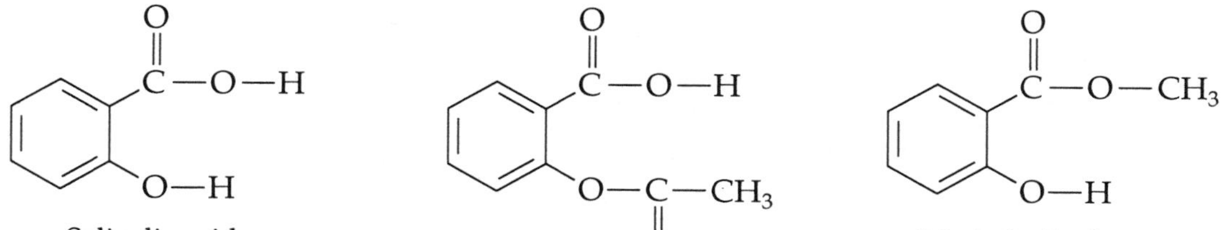

Salicylic acid
(antiseptic and wart remover)

Acetylsalicylic acid
(aspirin)

Methylsalicylate
(oil of wintergreen)

Figure 22.2   Some common salicylates.

PRE-LAB EXERCISE

1. Following the example below, calculate the aspirin concentration in mg/mL for each of the standard solutions that you will make in the first part of the procedure below.

2. Construct a data table in your notebook where you can record the concentrations and absorption values of your standard solutions.

3. Review the previous laboratory unit's description of the use of the Spectronic-20.

Example calculation:

The concentration of your stock standard aspirin solution is 0.10 mg/mL. Cuvette 2 contains 2.0 mL of standard aspirin solution + 8.0 mL HNO_3 + 1.0 mL iron(III) solution.

Milligrams of aspirin in cuvette 2 = (2.0 mL)(0.10 mg/mL) = 0.20 mg aspirin

Total volume = 2.0 + 8.0 + 1.0 = 11.0 mL

Concentration of aspirin in cuvette 2 = $\dfrac{0.20 \text{ mg}}{11.0 \text{ mL}}$ = 0.018 mg aspirin/mL

PROCEDURE

1. Preparation of Standard Aspirin Solutions

Your instructor will provide a stock standard aspirin solution. Record its concentration from the label. You will prepare five standards and a blank according to the table at the top of the next page. The blank contains everything that is in the standard solution except the colored species.

Use burets to measure the appropriate amounts of standard solution and nitric acid, and add these substances to each of six test tubes. Use a 1-mL pipet to add the iron solution to each test tube. Shake to mix thoroughly. Don't forget to label your test tubes. *Do not use the cuvettes as test tubes, and do not write on the cuvettes.*

Sample	Standard Solution (mL)	HNO$_3$ (mL)	Iron(III) 0.10 M (mL)
1 (blank)	0.0	10.0	0.0
2	2.0	8.0	1.0
3	4.0	6.0	1.0
4	6.0	4.0	1.0
5	8.0	2.0	1.0
6	10.0	0.0	1.0

Set the spectrophotometer wavelength to 535 nm. Refer to the previous laboratory unit for details regarding Spec-20 use. Keep one cuvette filled with your blank solution (sample 1). Use one or more other cuvettes to determine absorption of the standard solutions. After you analyze a sample, rinse the cuvette thoroughly with distilled water. Then rinse the cuvette once or twice with ~1 mL of the next sample before transferring the major portion of the solution whose absorption you will actually measure into the cuvette.

Zero the instrument using your blank (sample 1).

Transfer a volume of each standard solution to a cuvette and measure the absorption. Record this value in your notebook. Rezero the instrument using your blank between measurements. Your instructor may direct you to measure each standard solution more than once, rezeroing the instrument each time.

2. Analysis of Commercial Aspirin

After you have recorded the absorption values for all of the standard solutions, obtain two different samples of a commercial aspirin solution from your instructor. Record their numbers in your notebook.

For each commercial sample, pipet an 8.0-mL sample of the commercial solution into a large test tube. Pipet 2.0 mL of the 0.10 M HNO$_3$ solution and 1.0 mL of the iron solution into the test tube. Mix well.

Zero the spectrometer using your blank. Transfer a volume of each solution to a cuvette and measure and record the absorption.

Rinse all cuvettes several times with distilled water before returning them.

REPORT

1. Prepare a table that lists the concentrations (calculated in the pre-lab exercise) and corresponding absorption values for each of the standard aspirin solutions.

2. Either by hand or using a computer graphing program, make a calibration curve: plot the absorption values (y-axis) as a function of concentration (x-axis). Start from zero and draw the best straight line through your points.

3. On your graph, locate the absorption values for each commercial sample solution. From the point on the absorption axis (y-axis), draw a horizontal line to the calibration line.

Draw a perpendicular line from the calibration line to the x-axis to determine the concentration value.

4. Use the concentration value you determined in (3) above and information about the commercial solutions provided by your instructor to determine the number of milligrams of aspirin per tablet in the commercial products. How do your two samples compare with each other? How do they compare with the information on the product labels?

23

Introduction to Chromatography

OBJECTIVES

- Understand that mixtures must be separated in order that components might be studied separately.
- Observe and gain understanding of a physical separation technique, chromatography.
- Measure and understand the response factor (R_f) of different components in a mixture.
- Use chromatography and the R_f to identify an unknown and determine differences between fresh and cooked spinach.

EQUIPMENT/MATERIALS

Filter paper, ruler, pencil, toothpicks, beaker, 2% saline solution, aluminum foil or plastic wrap, food colors, rubbing alcohol, spinach, 95% ethanol, Kool-Aid® solutions.

INTRODUCTION

In order to study the chemical and physical properties of a particular element or compound, scientists routinely isolate their samples into pure substances. The presence of other substances in a sample affects the "behavior" of the substance being studied due to a variety of interactions. These interactions may disguise the very properties that the scientist (you) may be investigating. For example, if a biochemist wants to determine the behavior of a particular protein in the presence of an enzyme, he cannot have his protein sample cluttered with other proteins that might also interact with the enzyme and confuse his results.

A pure substance, at a given temperature, will be in a single phase (solid, liquid or gas) and is described as *homogeneous* (of one kind). Alternatively, mixtures fall into two categories; they can also be homogeneous, or they can be *heterogeneous* (of different kinds). Heterogeneous mixtures are usually obvious because different regions appear different to the naked eye. Heterogeneous mixtures can typically be separated by physical means quite easily. Examples of heterogeneous mixtures you might be familiar with include a pasta salad or a mixture of sand and water.

Homogeneous mixtures (called solutions) are hard to distinguish from pure substances because they cannot be identified by the naked eye. Indeed, they often cannot be identified even using a high-power microscope because the mixture exists at the molecular level. A well-stirred glass of sweetened ice tea with lemon is an example. The tea appears uniform to the naked eye, has a uniform density, uniform color and flavor, but could still be separated by physical means. For instance, the water (the most abundant substance) could be evaporated, and you would be left with a large number of other molecules including sugar, citric and ascorbic acid from the lemon, caffeine and tannins extracted from the tea leaves, minerals that were present in the water we used to brew the tea, and numerous other molecules.

Chemists often study macroscopic properties (e.g., boiling point, melting point, solubility, vapor pressure) to learn about molecular characteristics (e.g., molecular structure, bond angles, polarity). A critical aspect of any chemical investigation is to ensure that our observations truly reflect what we want to study. If we want to determine the temperature at which caffeine melts, it would not help us to melt a cup of coffee, even though we know that caffeine is present in that coffee. Somehow we need to isolate the caffeine as a pure substance before proceeding with our investigation.

In this experiment, you will use a technique called chromatography to separate and do some rudimentary analysis of different commercial dyes used in food coloring. You will also use paper chromatography to separate plant pigments. Chlorophyll is the green pigment in plants that absorbs light and helps the plant convert energy in photosynthesis. Dark-green leafy vegetables contain large amounts of chlorophyll. They also often contain carotene, a deep yellow-orange pigment that is a vitamin A precursor. Processing and extended storage cause destruction of the pigments in fruits and vegetables. The green color of water in which vegetables are heated comes from the release of chlorophylls due to the activity of the enzyme chlorophyllase. Chlorophyllase converts fat-soluble chlorophyll molecules to a water-soluble form collectively called chlorophyllides. If the heating is extreme or the chemical environment is acidic, the chlorophyll is converted to an olive-brown pigment that gives canned vegetables their characteristic color. In this experiment, you will use chromatography to separate the pigments in fresh and heat-processed spinach.

Chromatography

One of the most effective techniques for both identifying and separating mixtures is chromatography. There are many types of chromatography, but they all share the same basic theory. Chromatography involves the use of two "phases," one moving (the mobile phase) through the other (stationary phase). Often chromatography involves a tube that contains a column of a finely divided solid material (stationary phase) through which a solvent is moved (mobile phase) either by gravity or added pressure. The technique takes advantage of the fact that different components of a mixture interact differently with the two phases. So if you add your mixture to a chromatography system, some components will be more strongly attracted to (adsorbed onto) the stationary phase, while others will be more attracted to (soluble in) the mobile phase. As the mobile phase moves past the stationary phase, those components more strongly adsorbed to the stationary phase will lag behind their more mobile-phase-soluble counterparts, thereby effecting a separation. In other words, the substance that is more strongly attracted to the mobile phase will move rapidly with it, and the substance that is more strongly attracted to the stationary phase will move slowly and get left behind. Thus the two substances are separated.

Chromatography takes advantage of some kind of difference in properties between the mobile and stationary phases. For example, the mobile phase can be nonpolar (e.g., hexane or ethylacetate) and the stationary phase can be polar (silica gel, SiO_2). Conversely, the mobile phase can be polar and the stationary phase nonpolar. In a type of chromatography called ion exchange chromatography, the stationary phase contains ions with a particular charge, either positive or negative, that bind charged ions with opposite charges. Biochemists purifying proteins often use size exclusion as a way to separate substances. In this case, the stationary phase contains particles that are extremely porous. Very large molecules do not fit into the small pores and travel straight through the column with the mobile phase (which is usually an aqueous solution). Large molecules can penetrate into the pores a little bit and thus move more slowly through the stationary phase. Medium-sized molecules can travel deeper into the pores, moving slower still, and small molecules are the slowest of all, because they are able to diffuse into the smallest inner regions of the porous particles of the stationary phase.

The particular properties that enable chromatographic separations do not change for a particular substance from sample to sample as long as the chromatography conditions are the same. For example, suppose you use a column of silica (highly purified and finely ground sand or SiO_2) and a hexane solvent to separate substance X from substance Y. Substance Y drips off the column (elutes) first and substance X lags behind. Then when using silica with hexane, Y will *always* elute first and X will *always* follow behind. Furthermore, the amount or distance that they are separated will always be fairly consistent. We call this relative separation the *response factor* or R_f. The R_f can be defined as the distance traveled (or time required) by a particular substance divided by the distance traveled (or time required) by the solvent from the point at which the substance was loaded onto the stationary phase. A simpler version of R_f that relates particularly to the paper chromatography done in this experiment is shown below. Thus, R_f can be used to compare one chromatography to another. Two different substances might have the same R_f, but the same substance cannot have two different R_f's under the same conditions. Chromatography can therefore be used as a means of identification as well as separation.

In this laboratory, we will carry out a paper chromatography experiment to separate the dyes in food coloring using rubbing alcohol and saline solution as the mobile phase and a piece of a filter paper (cellulose) as the stationary phase. You will experiment with analysis and separation of the dyes in a commercial product. You will also separate pigments in fresh and heat-treated spinach. In all cases, you will determine the response factors of the different dyes and pigments. The formula for determining the R_f is

$$R_f = \frac{\text{Distance traveled by the dye}}{\text{Distance traveled by the solvent front}}$$

PRE-LAB EXERCISE

1. Suppose you ran two chromatographic experiments using the same dye. In the first experiment, you used a strip of paper that was 10 cm long, and in the second the strip was 20 cm long. Would you expect the R_f values to be the same or different? Briefly explain.

2. Suppose you ran two chromatographic experiments using the same dye. In the first experiment, you used water as a solvent, and in the second you used ethyl alcohol as the solvent. Would you expect the R_f values to be the same or different? Briefly explain.

3. Suppose you ran two chromatographic experiments using *different* dyes (same size paper and same solvent). Would you expect the R_f values to be the same or different? Briefly explain.

PROCEDURE

1. Analysis of Commercial Food Colors

Obtain a rectangular strip of the Whatman No.1 filter paper provided (approximately 11×19 cm). Use the ruler and a pencil to draw a line across the long end of the rectangle, approximately 1.5–2.0 cm from the bottom. (See Figure 23.1.) This will be the starting point for your chromatograph. Repeat this process for as many strips as you need. You will be able to test approximately six samples per strip.

Ensure that the beaker is taller than the strip of paper. (Trim the paper if necessary). Pour a small amount of 2% saline as solvent into the bottom. The depth of the solvent should be less than the 1.5–2.0 cm distance from the end of the paper to the starting point. (Thus, the starting line always remains above the level of the solvent.) You can run multiple chromatograms simultaneously if you use more than one beaker.

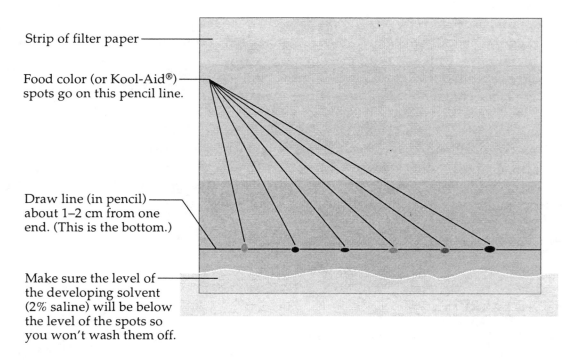

Strip of filter paper

Food color (or Kool-Aid®) spots go on this pencil line.

Draw line (in pencil) about 1–2 cm from one end. (This is the bottom.)

Make sure the level of the developing solvent (2% saline) will be below the level of the spots so you won't wash them off.

Figure 23.1   Spotting your filter paper with dye.

Obtain a spot plate or watch glass, and place a few drops of food coloring in each well.

Use a pencil to draw dots or crosshatches on the starting line of your strip of paper; space them 2–3 cm apart. Label these crosshatches with the different colors that you plan to use.

Use a separate toothpick for each color. Dip the end of the toothpick into the food color and carefully dot once onto the paper. Repeat for each color. Allow the paper to dry for at least 15 minutes.

When the paper is dry, roll it into a cylinder with the spots on the outside. Do not allow the two edges to touch, but staple it into a cylinder with a gap between the ends. (See Figure 23.2.)

Place the paper cylinder in the beaker, then gently cover the top of the beaker with foil or plastic wrap. (This allows the solvent to saturate the "atmosphere" inside the beaker and helps to pull the solvent up the strip of paper.)

Allow the solvent to move up the paper until it is 1–2 cm from the top edge of the cylinder. Do not allow the solvent to reach the top of the filter paper. Remove the cylinder and use a pencil to mark the solvent line at the top of the strip. Allow the cylinder to air dry.

After developing and removing the staples, your chromatogram should look something like the one in Figure 23.3.

After the strip has dried, mark the center of each colored dot. The center will be your best estimation as the spot will be uneven, as shown in the figure.

Measure from the start line to this center for the top value in your calculation of R_f for each of the dots. Measure from the start line to the solvent line. The latter is the bottom value for the R_f. Record these values in your notebook.

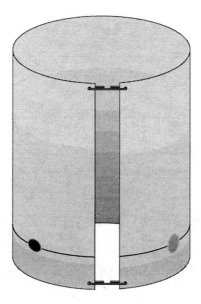

Paper is stapled so that the ends don't touch. Cylinder should be small enough that it doesn't touch the sides of the beaker.

Figure 23.2   Making a cylinder.

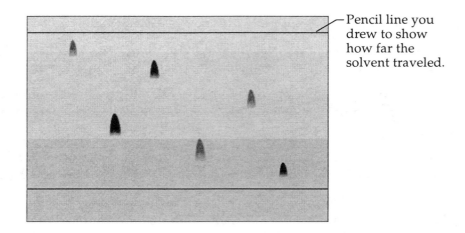

Figure 23.3   Chromatogram after developing and removing the staples.

Repeat this process until you have characterized all of the dyes in the commercial food-coloring package using 2% saline as a solvent. Pay particular attention to the dyes present in the green food coloring. Why is a green dye not listed as an ingredient in the package contents?

Repeat the analysis of the food coloring using *rubbing alcohol* (70% isopropyl alcohol) as a solvent. Use your results to decide what kind of solvent to use for the Kool-Aid® analysis and the unknown analysis below.

Repeat this process using the solutions of Kool-Aid® provided. If the chromatogram is too faint to read, you may need to repeatedly spot the paper. Your spot should be dark in color but not spread out over too much area. A concentrated spot works best, and this can be accomplished by waiting 5–10 minutes between spotting attempts to ensure that the Kool-Aid® solution has dried on the paper.

Record the artificial colors listed on the Kool-Aid® packages in your notebook.

Repeat this process, analyzing two of the unknowns provided by your instructor.

2. Separating Plant Pigments

Rinse torn spinach leaves in water. Separate the leaves, and place half of them in a beaker with a small amount of water. Heat on a hot plate just until the water becomes green. Cool and drain.

Pour a small amount of ethyl alcohol into the bottom of your beaker.

On a piece of filter paper, draw a line in pencil 1 cm from the bottom.

Make spots of cooked and uncooked spinach on the line by pressing a toothpick gently against the leaf and onto the paper so that a small spot appears. Reapply the spinach to the spots several times, allowing the spots to dry between applications. The spots should be small; otherwise, a streak will be obtained instead of an actual separation.

Prepare the cylinder as before. Place it in the beaker and cover with aluminum foil or plastic wrap. Make sure that the solvent is below the pencil line and sample spots and that the paper is not touching the sides of the chamber. Allow the solvent to rise up through the paper until it is about 1 cm from the top of the paper.

Remove the paper and mark the solvent front line with a pencil. Also use the pencil to circle any spots that have separated. Allow the paper to dry. Measure and record the distance traveled by the solvent front and any of the spots you detected. Determine their R_f values.

REPORT

1. Construct four tables to record the data from each of your food-coloring chromatograms. An example is given below. Each table should be properly labeled for the solvent used and the colors tested. If more than one dye is present in a given color, list each dye color and the corresponding R_f value separately.

Chromatograph #_____   Solvent_____

Food color	Distance traveled by dye(s)	Distance traveled by solvent	R_f value

2. Which solvent did you use for the final two separations? What were the experimental factors behind your decision? How did the R_f values differ in the two solvents? Which is a better solvent for analysis of these food colorings? Briefly explain.

3. Compare the R_f values for the dyes found in the Kool-Aid® to those you tested in part 1. Do any of the Kool-Aid® samples contain the same dyes as the food coloring? Do any of the Kool-Aid® samples contain the same dyes as other Kool-Aid®? Compare your experimental results with the information provided on the labels.

4. Report on your analysis of the unknown(s). Which dye(s) did it/they contain?

5. Report on your analysis of the spinach. Make a table that lists pigment color and R_f value for each of the two samples. Postulate as to the identity of the spots given their color and relative R_f values. (Your instructor may wish to give you additional information about the relative polarities of the different pigments.) How did heating affect the pigment content of the spinach?

24

Fermentation and Distillation

OBJECTIVES

- Use yeast and sugar to effect a fermentation.
- Purify the alcohol obtained by fermentation through simple distillation.
- Compare densities of liquid before and after fermentation (and with commercial distilled spirits).

EQUIPMENT/MATERIALS

Materials for fermentation (week 1): Dry baker's yeast, sugar, water, flask, test tube, paraffin or vegetable oil, one-hole stopper, glass and rubber tubing, calcium hydroxide solution (optional).

Materials for distillation (week 2): Distillation apparatus, beakers, flasks, graduated cylinders, commercial distilled spirits.

INTRODUCTION

Fermentation

Fermentation is the breakdown of complex molecules to simpler ones through the action of some microorganism, such as yeast. Indeed, we normally think of fermentation in the context of beer or wine, but the term also includes the conversion of milk to yogurt or cheese and the making of bread. In all of these examples, carbohydrates are being fermented by yeast. In the case of bread, yeast converts carbohydrates into carbon dioxide gas (CO_2). The little pockets of CO_2 make the bread rise and produce the fluffy consistency that we associate with bread. It is this formation of gas pockets that distinguishes bread from crackers. For fermentations that produce alcohol, such as wine, carbohydrate is converted to ethanol and carbon dioxide, as summarized in the following equation:

$$C_6H_{12}O_6 \xrightarrow{\text{Yeast}} 2\,C_2H_5OH + 2\,CO_2$$

Glucose   Ethanol   Carbon dioxide

Yeast is naturally present on the skin of grapes, which is why the making of wine dates back before recorded history. Current professional wine makers kill the natural yeast and replace it with commercial yeast to better control the fermentation process. This process must be done in the absence of air (specifically oxygen), lest the ethanol be oxidized to acetic acid, which is the chemical we associate with vinegar:

$$C_2H_5OH \xrightarrow{\text{Air oxidation}} HC_2O_2H_3$$
$$\text{Ethanol} \qquad\qquad\qquad \text{Acetic acid}$$

Fermentation continues until the ethanol concentration is high enough to inhibit yeast growth. This usually happens when the alcohol content is between 10 and 14%. If you want to produce a beverage that has a greater alcohol content, you must either distill it or fortify it with alcohol that resulted from the distillation of another product of fermentation. Whiskeys, vodka, and gin are examples of distilled spirits. Port wine, cognac, and liqueurs are examples of fortified wines.

Distillation

Distillation is a purification process wherein a substance is heated to its boiling point, the vapor produced upon boiling is allowed to flow away from the boiling liquid, and the vapor is cooled to condense it back to the liquid. (See Figure 24.1.) What is the advantage of converting a liquid to its vapor, moving it to a different place, and reconverting it to the liquid form? The advantage depends on the differences between a liquid and its vapor. Suppose you have a liquid that contains contaminants, like seawater. You cannot drink seawater because of its high salt content. Indeed, drinking seawater has the opposite of the desired consequence of relieving thirst: you wind up more dehydrated than before because

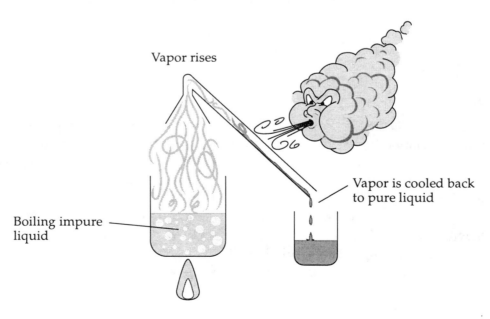

Figure 24.1   How a distillation works.

your body has to excrete additional water to balance the high levels of salt in the seawater. Now suppose you boil (or evaporate) the seawater. This vapor phase will contain almost pure water. The nonvolatile salts will not vaporize at 100°C and will be left behind. Thus if you evaporate a container of seawater, the salt will be left behind, clinging to the walls and bottom of the container. If you can capture that water vapor somehow—say by moving it to a place where you can cool and condense it—you will have water that is pure enough to drink. Remember this the next time you are stranded on a desert island!

Distillation is a common way to separate and purify crude oil into its usable components (gasoline, kerosene, asphalt, etc.). As mentioned above, distillation also enables one to produce a beverage that has a higher alcohol content than that obtained from a fermentation process. The fermentation process leaves behind a variety of substances besides ethanol: yeast, unfermented sugar, residue from the plants that were fermented, water, and so on. Distillation will separate the alcohol from all of those contaminants. It also has the effect of concentrating the alcohol. The maximum concentration of alcohol that can be obtained through distillation from a water-based fermentation is 95%. This is because ethanol and water form a mixture (called an azeotrope) that boils at a constant composition. Thus, the vapor that forms upon the boiling of any water–alcohol mixture is 95% ethanol and 5% water.

What is meant by the term *proof*? The proof of an alcoholic beverage is twice the volume percent of ethanol. This means that if you have 80-proof vodka, the solution is 40% ethanol and 60% other things (mostly water) by volume. The word *proof* comes from an old technique of measuring the purity of distilled spirits. A sample of distilled spirits was poured over gunpowder and the mixture was ignited. If the alcohol had a volume percent of 50 or more, one would observe a characteristic blue flame. Thus if it were 50% alcohol, it was 100% proved: 100-proof. Nowadays we use simple density measurements to determine the alcohol percentage.

In this experiment, you will ferment a sugar–water mixture using ordinary baker's yeast. You will then do a simple distillation of the fermentation mixture to obtain a purified ethanol. You will determine its proof and perhaps compare it to the proof of a sample of commercially available alcohol.

PRE-LAB QUESTIONS

1. Prepare your notebook for this experiment.
2. Calculate the density for each of the following volume/mass measurements of different alcohol–water mixtures. Postulate as to which samples might have the same proof.

Sample	Volume (mL)	Mass (g)
A	2.6	2.29
B	8.1	6.32
C	5.3	4.66
D	3.9	3.67
E	4.7	3.67
F	7.2	6.77
G	1.1	0.97

PROCEDURE

Week 1: Fermentation

Obtain a 250-mL Erlenmeyer flask. Weigh out 20 g of sugar and place it in the flask. Add 85 mL of water and swirl the flask gently until the sugar is thoroughly dissolved.

Add 1.0 g of dry baker's yeast. Gently swirl the flask for an additional 30 seconds.

Place a one-hole rubber stopper that is fitted with a small length of glass tubing into the neck of the fermentation flask. (See Figure 24.2.) Attach a length of rubber tubing to the glass tubing. Attach another short length of glass tubing or a Pasteur pipet to the other end of the rubber tubing. The latter goes into a test tube containing either ordinary water or a solution of $Ca(OH)_2$. Cover the top of the test tube's aqueous contents with a 1-cm layer of paraffin oil. (Ordinary vegetable oil may be substituted.) This setup will exclude air from the system.

Label your fermentation setup with your name and lab section, and place it somewhere where it won't be disturbed or exposed to extreme heat or extreme cold. Fermentation needs to proceed for 5–7 days.

How the air exclusion system works: When the yeast fermentation process begins, carbon dioxide is generated (see the reaction chemistry above). If this process were done in a sealed system, pressure would begin to build up and would eventually blow off the stopper. The air exclusion system allows the gas to be vented to the outside without allowing air back in. The gas moves through the tubing and bubbles up through the water and through the oil, but the oil and water prevent any air from entering the system. The oil

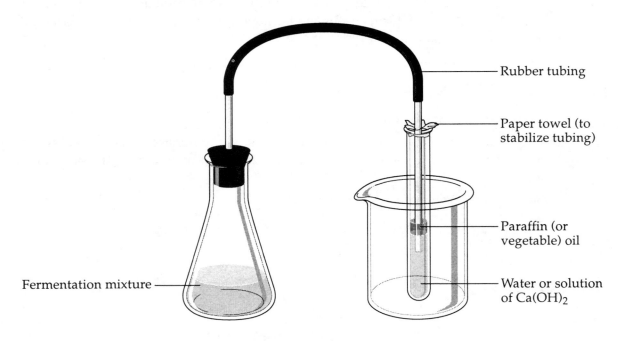

Fermentation mixture

Rubber tubing

Paper towel (to stabilize tubing)

Paraffin (or vegetable) oil

Water or solution of $Ca(OH)_2$

Figure 24.2   Fermentation setup.

also prevents the water from evaporating through the course of the week. If you use a solution of $Ca(OH)_2$, the solution will turn cloudy as the CO_2 generated reacts with the calcium hydroxide:

$$Ca(OH)_2 + CO_2 \longrightarrow CaCO_3 \text{ (white solid)} + H_2O$$

You may actually see the CO_2 bubbles forming and moving through the liquid, but you may not. Do not be concerned if you don't see bubbles; chances are good that fermentation is proceeding nonetheless.

Week 2: Distillation

Avoid shaking the fermentation flask and disturbing the sediment. Remove the rubber stopper and tubing from your fermentation flask. Rinse the tubing with water to remove the white solid adhering to it and return it to the equipment bench. Your instructor will direct you how to discard the liquid in the air exclusion apparatus.

Carefully decant the fermentation mixture into a clean 250-mL Erlenmeyer flask. Try to leave as much sediment as possible in the flask and pour only the liquid portion. Use a pipet to remove a small volume. Determine the density of this small portion by accurately measuring the mass and volume. Record these values in your notebook.

Slowly heat the contents of the flask to ~65°C, stop heating, and allow the flask to cool. This stops the fermentation process and keeps the contents of the flask from foaming during distillation.

Once the flask has cooled, assemble the distillation apparatus shown in Figure 24.3 on the next page. Insert a one-hole stopper equipped with a length of glass tubing into the flask. The test tube that will be collecting the distillate should be well secured by a clamp attached to the ring stand. Be certain that the end of the glass tubing does not touch the bottom of the test tube. Expose as much of the outside surface area of the collection test tube to the ice water bath as possible to maximize condensation.

Slowly and gently heat the flask contents until they begin to boil. Control the heat to maintain smooth and steady boiling. You should be able to see vapor moving through the glass tubing and condensing in the test tube. Continue heating until you have collected about 6 mL of distillate (approximately one-third of the test tube).

Turn off the heat source, and carefully remove the glass tubing from the test tube.

Weigh a clean, dry 10-mL graduated cylinder. Pour approximately 4 mL of your distillate into the graduated cylinder, but record the exact volume. Weigh the graduated cylinder and its contents and determine the mass of the contents. Determine the density of the distillate (density = mass/volume).

Repeat the density determination described above for one or more commercial samples of distilled spirits as directed by your instructor.

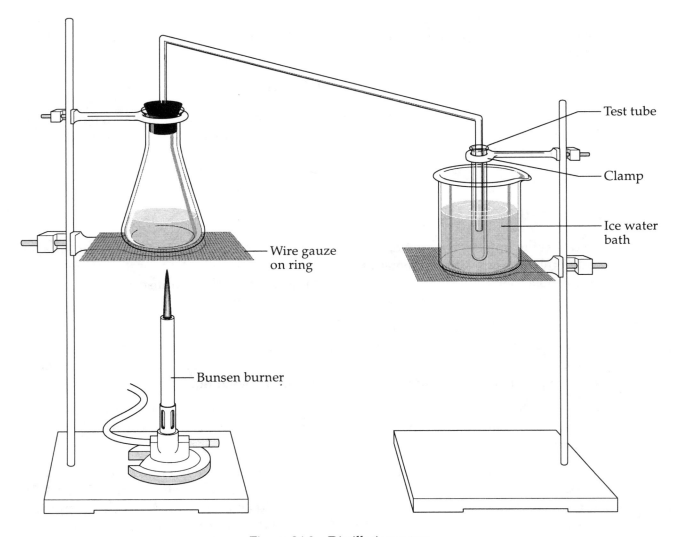

Figure 24.3   Distillation setup.

REPORT

1. Briefly outline your fermentation and distillation procedure. Include any specific observations that you made during the process.

2. Prepare a table that lists the masses, volumes, and calculated densities of the liquid samples before distillation and after distillation. What did the distillation accomplish?

3. In a table or paragraph, report the results of your comparison of the density of your distillate and the commercial distilled spirit(s).

4. What conclusions can you draw about your distillation process based on your density results?

25

Polymerization Reactions

OBJECTIVES

- Investigate different polymerization processes.
- Prepare polymers and compare their properties.

EQUIPMENT/MATERIALS

Styrene, t-butyl peroxide benzoate, xylene, polystyrene packing materials, solutions of hexamethylene diamine, adipoyl chloride and sodium hydroxide, beakers, tongs, aluminum foil.

INTRODUCTION

Polymers are giant molecules made of repeating smaller units. The starting material, which is a single unit, is called a monomer. Many important biological compounds are polymers. Cellulose (cotton, wood) and starch are polymers of glucose units; proteins are polymers of amino acids, and nucleic acids are polymers of nucleotides. Since the 1930s, a large number of man-made polymers have been manufactured. The simplest of these is polyethylene; you may be familiar with this polymer as clear plastic wrap or the transparent plastic bags you use to buy produce at the grocery store.

Polyethylene is synthesized by chemically linking together individual ethylene units until you finally have a single enormous molecule in the form of a long chain, as shown in Figure 25.1. This process is called an *addition polymerization* because two mono-mers are simply added to each other with the loss of the double bond. However, this reaction will not occur if you simply mix ethylene molecules together; the polymerization process requires the help of an unstable molecule called an *initiator*. Benzoyl peroxide and t-butyl peroxide benzoate are examples of initiators.

Benzoyl peroxide contains an oxygen–oxygen single bond, which ruptures when exposed to heat or ultraviolet light to yield two halves, called free radicals. A free radical is

---

$$H_2C{=}CH_2 + H_2C{=}CH_2 \longrightarrow -CH_2-CH_2-CH_2-CH_2-$$

$$-CH_2-CH_2-CH_2-CH_2- + H_2C{=}CH_2 \longrightarrow -CH_2-CH_2-CH_2-CH_2-CH_2-CH_2-$$

$$\Big\downarrow$$

$$\cdots -CH_2-CH_2-CH_2-CH_2-CH_2-CH_2-CH_2-CH_2-CH_2-CH_2-CH_2-CH_2- \cdots$$

Figure 25.1   Synthesis of polyethylene from individual ethylene units.

a molecular fragment that has one unpaired electron (designated in Figure 25.2 by the unpaired dot). Free radicals will often initiate polymerization reactions. Figure 25.3 shows a free-radical initiator reacting with an ethylene molecule to produce another free radical. The new free radical in turn reacts with another ethylene molecule, and the chain grows longer and longer until two free radicals collide with each other and the process stops. Because initiators are unstable compounds, care should be taken to keep them away from flames or direct heat. Even dropping the bottle containing a peroxide initiator may create a minor explosion, so be very careful.

In this experiment, you will synthesize polystyrene from styrene (Figure 25.4). Note the similarity between polyethylene and polystyrene. What is different about the two polymers? Polystyrene has two familiar macroscopic forms. One is the foam form (one brand of which is Styrofoam®) in which the polymer is extruded with a gaseous propellant. This form is used in foam coffee cups and packing peanuts. The other is a more brittle plastic form used often to package food; the container that you might use for a "to-go" salad from a supermarket salad bar is likely to be made of polystyrene. Although the two substances appear very different to us, at the molecular level, the polymers are the same. The recycling code for polystyrene is #6 PS, although at this time, its recycled uses are very few.

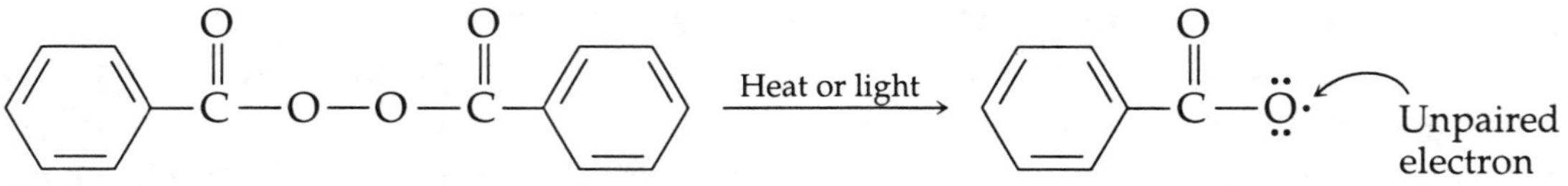

Benzoyl peroxide   Free radical

Figure 25.2   Formation of a free radical initiator from benzoyl peroxide.

$$R{-}O{\cdot} + H_2C{=}CH_2 \longrightarrow R{-}O{-}CH_2-\overset{\cdot}{C}H_2$$

$$R{-}O{-}CH_2-\overset{\cdot}{C}H_2 + H_2C{=}CH_2 \longrightarrow R{-}O{-}CH_2-CH_2-CH_2-\overset{\cdot}{C}H_2$$

$$R{-}O{-}CH_2-CH_2-CH_2-\overset{\cdot}{C}H_2 + H_2C{=}CH_2 \longrightarrow$$

$$R{-}O{-}CH_2-CH_2-CH_2-CH_2-CH_2-\overset{\cdot}{C}H_2$$

Figure 25.3   Free-radical initiator of ethylene.

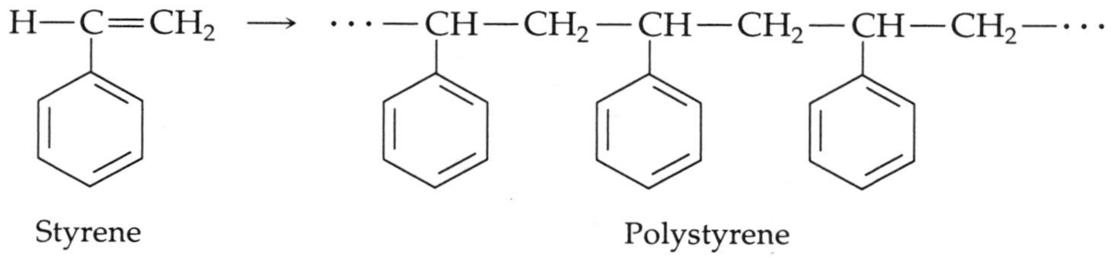

Figure 25.4 Synthesis of polystyrene from styrene.

The second type of polymerization reaction in this experiment is an example of a *condensation polymerization*, so-called because two monomers are condensed into a longer unit and at the same time a small molecule is eliminated. Figure 25.5 shows nylon 66 being made by condensing adipoyl chloride and hexamethylene diamine. This polymer is called nylon 66 because there are six carbon atoms in the adipoyl chloride and six carbon atoms in the diamine. Nylon 6-10 is made from hexamethylene diamine and sebacoyl chloride (containing ten carbons atoms). NaOH is added to the polymerization reaction mixture in order to neutralize the HCl that is released.

The hexamethylene diamine is water-soluble. The adipoyl chloride is dissolved in cyclohexane. The two solvents are not miscible, meaning that they will not dissolve in each other but will form two layers. The nylon is formed at the interface where the two layers come in contact with each other (Figure 25.6).

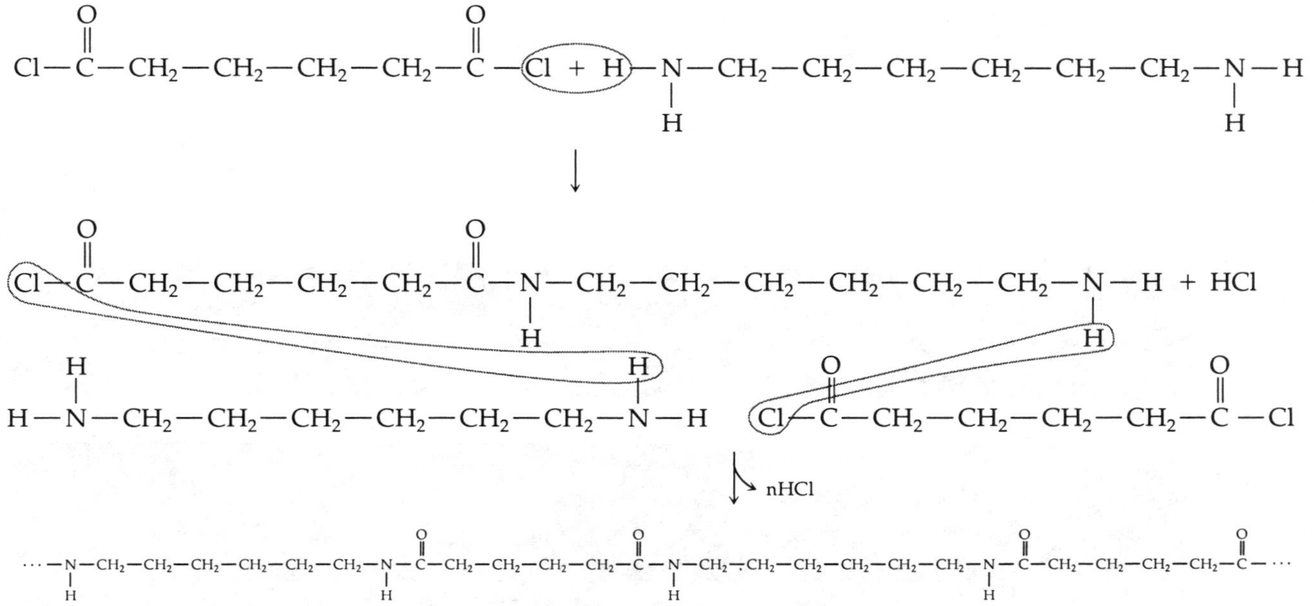

Figure 25.5 Adipoyl chloride condensing with hexamethylene diamine to make nylon.

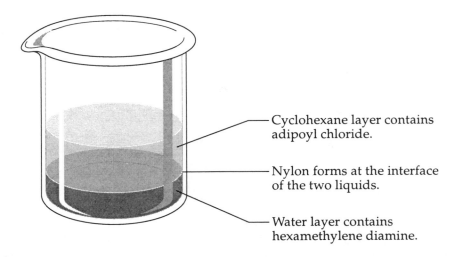

Figure 25.6 Nylon-66 forming at the interface of water and cyclohexane.

PRE-LAB EXERCISE

1. Write the addition polymerization reaction of vinyl chloride, $CH_2=CHCl$, to make polyvinyl chloride (PVC).

2. Write the condensation polymerization reaction for the formation of the polyester made from ethylene glycol ($HO–CH_2–CH_2–OH$) and phthalic acid. What is the condensation product?

Phthalic acid

PROCEDURE

CAUTION! POLYSTYRENE, ADIPOYL CHLORIDE, HEXAMETHYLENE DIAMINE, CYCLOHEXANE, AND XYLENE ARE ALL EXTREMELY FLAMMABLE. DO NOT USE A FLAME FOR ANY PURPOSE IN THIS EXPERIMENT.

CAUTION! PEROXIDES ARE UNSTABLE AND SHOULD NEITHER BE HEATED DIRECTLY NOR DROPPED. ADIPOYL CHLORIDE AND 80% FORMIC ACID ARE CORROSIVE AND WILL IRRITATE YOUR SKIN ON CONTACT. ALL OF THE CHEMICALS USED IN THIS EXPERIMENT ARE TOXIC. USE CAUTION IN HANDLING THEM AND WASH YOUR HANDS THOROUGHLY BEFORE LEAVING THE LABORATORY.

Synthesis of Polystyrene

Place 25 mL of styrene in a 150-mL beaker. Add 20 drops of the t-butyl peroxide benzoate initiator. Mix the solution by swirling. Warm the beaker on a hot plate in the hood to approximately 140°C. Record any observations.

When bubbles appear, remove the beaker from the hot plate with tongs. The polymerization reaction is exothermic. Overheating could create sudden boiling. When the bubbles disappear, put the beaker back on the hot plate, but every time the mixture starts boiling, remove the beaker. Continue heating until the mixture has a thick syrupy consistency.

Pour the mixture onto a square of aluminum foil and let it solidify.

The residual polystyrene in the beaker can be cleaned out by adding xylene and warming it on the hot plate in the hood until the polymer is dissolved. Pour a few drops of the warm xylene solution on another piece of aluminum foil and let the solvent evaporate. A thin film of polystyrene will be obtained. This is one of the techniques, called solvent casting, used to make films from bulk polymers. Discard the remaining xylene solution into the appropriate waste container. You can wash your beaker with soap and water.

Investigate the consistency of the solidified polystyrene on the aluminum foil. You can remove the solid mass by prying it off with a spatula. Obtain a piece of polystyrene packing material and compare it to the polystyrene that you synthesized. Place the piece of packing material on another piece of foil and add a few drops of acetone to it. Describe what happens and compare the result to the polystyrene that you synthesized.

Synthesis of Nylon 66

Place 10 mL of a 5% aqueous solution of hexamethylene diamine into a 50-mL beaker. Add 20 drops of a 30% NaOH solution. Use a stirring rod to stir briefly.

Obtain 10 mL of a 5% adipoyl chloride solution in cyclohexane. Tilt the beaker containing the hexamethylene diamine to approximately 45° and carefully pour the adipoyl chloride solution down the side of the beaker so as to layer it on top of the aqueous solution below without disturbing it too much. The nylon will form at the interface of the two liquids.

With the bent copper wire provided, slowly lift the film from the center of the interface. Pull it slowly out of the beaker and wrap it around a wooden stick.

CAUTION: DO NOT TOUCH THE NYLON ROPE WITH YOUR HANDS.

Experiment with how you pull out the nylon; pull it fast or slow and see what effect it has on the quality of the rope. If it breaks, simply begin again at the center of the interface. If you get bored with making rope, you can stir the mixture vigorously to make a large nylon mass.

Rinse the nylon rope off thoroughly with water; once the nylon has been rinsed, you may safely touch it with your hands. Dissolve a small amount of the nylon in 80% formic acid. Place a few drops of the solution onto a microscope slide and evaporate the solvent in the hood. Compare the appearance of the solvent-cast nylon film with that of the polystyrene.

**DO NOT DISPOSE OF ANY OF THE NYLON PRODUCTS IN THE SINKS UNLESS YOU
ARE A LICENSED PLUMBER AND ARE PREPARED TO UNCLOG THE DRAINS!**

You may discard the washed nylon rope and mass into the trash can. Any leftover nylon solutions should be discarded into the waste containers provided for that purpose.

REPORT

1. Write a brief paragraph or two describing the procedure and your observations. Summarize your results in another brief paragraph or two. Compare and contrast the two polymers you made (according to solubility, strength, appearance, synthesis, solvent casting, and so on). Compare the polymers you made with similar versions that are commercially available.

2. Write the structure of the monomers and that of the repeating unit in nylon 10-6.

3. A polyester is made of adipoyl chloride and ethylene glycol. Draw its structure. What molecule has been eliminated in this condensation reaction?

4. Polymethylmethacrylate is an addition polymer commonly known as Lucite® and Plexiglas®. It is made from the monomer methylmethacrylate. Draw the structure of polymethylmethacrylate.

$$H_2C = C\begin{array}{c} \\ \\ \end{array}\begin{array}{c} C-O-CH_3 \\ \| \\ O \end{array}$$
CH₃

Methylmethacrylate

26

Chemiluminescence and Antioxidants

OBJECTIVES

- Qualitatively explore a chemiluminescent reaction.
- Observe the effects of antioxidants on the rate of a chemiluminescent reaction.

EQUIPMENT/MATERIALS

Test tubes, graduated cylinder, solution of luminol, potassium carbonate, and copper sulfate, other solutions of hydrogen peroxide, antioxidant substances, and oxidizing agents.

INTRODUCTION

Chemiluminescence

If you have ever seen the glittery display of fireflies on a summer evening or the eerie nighttime glow produced by the wake of a boat or crashing waves in the ocean, you have observed a chemical reaction that is releasing energy. Many chemical reactions produce energy, typically in the form of heat. However, a small class of reactions produces energy in the form of light; these reactions are called *chemiluminescent* reactions. Man-made versions include the glow sticks or lightsticks that are sold for camping, novelty, and roadside emergencies. The biological version of this process (e.g., fireflies, glowworms) is called bioluminescence.

Chemiluminescent reactions typically involve the oxidation of a compound. The oxidized form is produced in a chemically excited state of higher energy. As the compound returns to a more stable form, it emits its extra energy as light. (See Figure 26.1.) The color of the light depends on the energy gap between the excited state and the lower-energy ground state. Chemiluminescence can also result when energy from an excited molecule is transferred to another molecule. This second molecule then emits the light energy as it returns to its ground state. (See Figure 26.2.) Most commercially available lightsticks use the latter process.

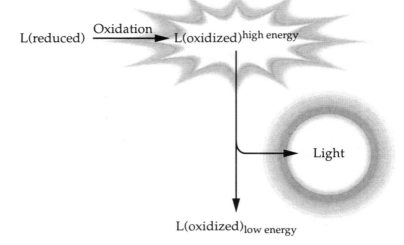

Figure 26.1   Schematic representation of the chemiluminescent reaction of molecule L. The oxidized form of L is produced in a high-energy excited state. As it returns to the more stable ground state, it gets rid of the extra energy as light.

Commercial lightsticks typically use dilute hydrogen peroxide to oxidize a phthallic ester in the presence of a dye. The high-energy oxidation product transfers its energy to the dye molecule, which then emits light, as illustrated in Figure 26.2. Different dyes produce different colors. The lightstick is constructed of an outer thick plastic tube around a smaller inner tube made of glass, which contains the hydrogen peroxide. The inner tube is cracked, the two chemicals are mixed, and the reaction occurs.

In this experiment, you will be exploring the chemiluminescent reaction of a compound called *luminol*. The less charming and more technical name for luminol is 3-amino-phthalhydrazide, and its structure and oxidation are shown in Figure 26.3. In an alkaline solution with a metal catalyst, luminol can be oxidized by hydrogen peroxide. Law enforcement agencies use luminol to detect blood at crime scenes; the iron in blood can catalyze

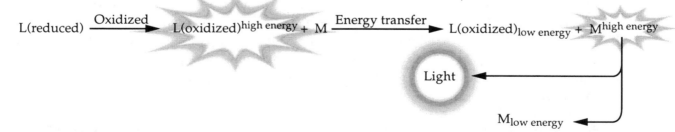

Figure 26.2   In this example the energy is transferred from the oxidized form of L to a different molecule, M. M then returns to a more stable, lower energy form and emits light in the process. The light released in this case is often a different color from that in the example in which the oxidized form of L emits light directly.

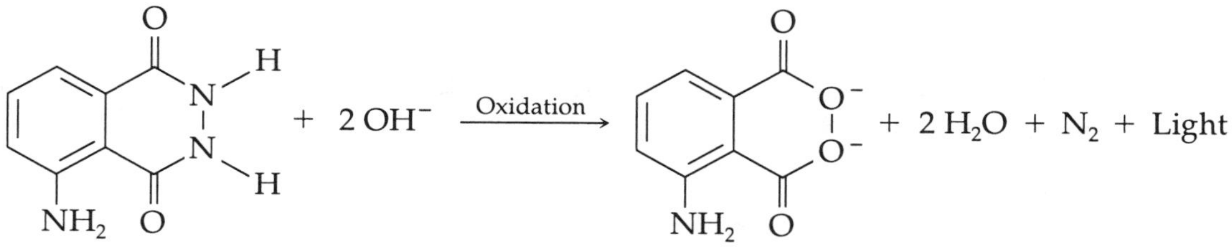

Figure 26.3   Oxidation of luminol (3-aminophthalhydrazide).

the reaction. Because this is an oxidative process, the reaction is enhanced by the presence of oxidizing agents and inhibited by the presence of antioxidants.

Antioxidants

Antioxidants are a class of chemical compounds that are gaining increased media coverage for their importance in healthy cellular processes. They help to rid the body's cells of dangerous "free radicals," which, if left unchecked, can promote oxidative processes that harm the cell. Free radicals are a natural by-product of normal metabolism, and your body has several mechanisms for dealing with them. However, antioxidants such as vitamins A, C, and E can also assist your cells in protecting against free radical damage. Ascorbic acid, or Vitamin C, is an important water-soluble antioxidant that you'll use in this experiment.

If an antioxidant is present, oxidative reactions are inhibited. This means that they either don't occur at all, or occur more slowly than if the antioxidant weren't present. Other compounds enhance oxidation processes. Because the oxidative process that you are studying is a chemiluminescent reaction and gives off light, you'll be able to detect the presence and absence of an antioxidant by how much the reaction mixture glows. In this experiment, you will mix two solutions and observe how intensely the solution glows. This is a qualitative comparison, and subjectivity may make comparisons difficult. Do the best you can, and always prepare your control solution at the same time as your test solution.

PRE-LAB EXERCISE

1. Think about what you may have read about the substances in the procedures that act as antioxidants or oxidants. Look them up if necessary. Which substances would you predict to enhance the oxidation of luminol and which would inhibit it?

2. Why is it important to prepare a control reaction for every single trial, rather than just following the usual protocol and performing one in the beginning?

3. Why is it necessary to have potassium carbonate (K_2CO_3) present in the reaction solution?

PROCEDURE

Bright light will make comparisons difficult. Lights should be dimmed if possible, but not turned so low as to make the laboratory unsafe.

Solution 1 contains luminol, potassium carbonate (K_2CO_3), and the metal catalyst copper sulfate, in the following concentrations:

Luminol	4.0×10^{-3} M
K_2CO_3	0.15 M
$CuSO_4$	3.5×10^{-3} M

Solution 2 contains hydrogen peroxide in the following concentration:

H_2O_2	4.5×10^{-2} M

In addition, you will use several solutions of oxidants and antioxidants.

You should always prepare a control reaction for comparison to your test reaction.

To prepare a control reaction, obtain a small test tube. Combine 2 mL of solution 1 with 2 mL of distilled water and stir. Add 2.0 mL of solution 2. Observe the reaction. You may wish to time how long the reaction lasts.

Compare the control reaction to that of each of the mixtures described in the table below, and record your observations. In each case, combine the antioxidant solution with solution 1 and ensure complete mixing before adding solution 2. Try to add solution 2 to both the control reaction and the test reaction at the same time.

Trial	Solution 1	Water	Solution 2	Oxidant/Antioxidant
Control	2.0 mL	2.0 mL	2.0 mL	None
1	2.0 mL	None	2.0 mL	2.0 mL ascorbic acid (vitamin C) solution
2	2.0 mL	None	2.0 mL	2.0 mL sucrose solution
3	2.0 mL	None	2.0 mL	2.0 mL methionine solution
4	2.0 mL	None	2.0 mL	2.0 mL propylgallate solution
5	2.0 mL	None	2.0 mL	2.0 mL sodium thiosulfate solution
6	2.0 mL	None	2.0 mL	2.0 mL green tea
7	2.0 mL	None	2.0 mL	2.0 mL NaOCl (bleach) solution
8	2.0 mL	None	2.0 mL	2.0 mL bovine hemoglobin solution
9–?	2.0 mL	None	2.0 mL	2.0 mL fresh fruit or vegetable juice[*]

[*]Options to try: grape juice, raspberry or blueberry juice, carrot juice. You should either obtain unpasteurized juice from a health food store or squeeze it fresh from the fruit or vegetable yourself. Comparing unpasteurized juices to pasteurized juices is an interesting exercise. Note that exceptionally acidic juices, such as citrus juices, may lower the pH below 7. The reaction is inhibited at low pH. One can control for pH changes by buffering the fruit or vegetable juice with $Na_2CO_3/NaHCO_3$ or using a citrate buffer. Excess sugar that is often added to juices may enhance the reaction. Students may also wish to explore other foods that are touted as antioxidants in the media (e.g., garlic, onion, broccoli). Note that oil-soluble antioxidants (e.g., vitamin E or vitamin A) do not work well as inhibitors in this reaction due to solubility problems.

REPORT

1. Prepare a table that reports the results of the oxidation of luminol in the presence of different antioxidant/oxidant substances. Are you able to determine whether something has antioxidant properties from the results of your experiment? How close were your predictions to the results you obtained? Are you able to account for any discrepancies?

2. Review Beer's law and spectrometry (Units 21 and 22) and design an experiment in which you use this reaction to obtain quantitative information about the concentration of antioxidants in foods.